D0060870

ENGINEERING SOCIETIES MONOGRAPHS

FOUR national engineering societies, the American Society of Civil Engineers, American Institute of Mining and Metallurgical Engineers, The American Society of Mechanical Engineers, and American Institute of Electrical Engineers, have made arrangements with the McGraw-Hill Book Company, Inc., for the production of selected books adjudged to possess usefulness for engineers or industry, but not likely to be published commercially because of too limited sale without special introduction. The societies assume no responsibility for any statements made in these books. Each book before publication has, however, been examined by one or more representatives of the societies competent to express an opinion on the merits of the manuscript.

ENGINEERING SOCIETIES MONOGRAPHS COMMITTEE

Engineering Societies Library,
New York

Applied
Hydro- and Aeromechanics

BASED ON LECTURES OF
L. Prandtl, Ph. D.

BY
O. G. Tietjens, Ph.D.

TRANSLATED BY
J. P. Den Hartog, Ph.D.

DOVER PUBLICATIONS, INC.
NEW YORK • NEW YORK

Standard Book Number: 486-60375-X
Library of Congress Catalog Card Number: 57-4858

Manufactured in the United States by Courier Corporation
60375X17
www.doverpublications.com

ENGINEERING SOCIETIES MONOGRAPHS

For many years those who have been interested in the publication of papers, articles, and books devoted to engineering topics have been impressed with the number of important technical manuscripts which have proved too extensive, on the one hand, for publication in the periodicals or proceedings of engineering societies or in other journals, and of too specialized a character, on the other hand, to justify ordinary commercial publication in book form.

No adequate funds or other means of publication have been provided in the engineering field for making these works available. In other branches of science, certain outlets for comparable treatises have been available, and besides, the presses of several universities have been able to take care of a considerable number of scholarly publications in the various branches of pure and applied science.

Experience has demonstrated the value of proper introduction and sponsorship for such books. To this end, four national engineering societies, the American Society of Civil Engineers, American Institute of Mining and Metallurgical Engineers, The American Society of Mechanical Engineers, and American Institute of Electrical Engineers, have made arrangements with the McGraw-Hill Book Company, Inc., for the production of a series of selected books adjudged to possess usefulness for engineers or industry but of limited possibilities of distribution without special introduction.

The series is to be known as "Engineering Societies Monographs." It will be produced under the editorial supervision of a Committee consisting of the Director of the Engineering Societies Library, Chairman, and two representatives appointed by each of the four societies named above.

Engineering Societies Library will share in any profits made from publishing the Monographs; but the main interest of the societies is service to their members and the public. With their aid the publisher is willing to adventure the production and distribution of selected books that would otherwise be commercially unpractical.

Engineering Societies Monographs will not be a series in the common use of that term. Physically the volumes will have similarity, but there will be no regular interval in publication, nor relation or continuity in subject matter. What books are printed and when will, by the nature of the enterprise, depend upon the manuscripts that are offered and the Committee's estimation of their usefulness. The aim is to make accessible to many users of engineering books information which otherwise would be long delayed in reaching more than a few in the wide domains of engineering.

ENGINEERING SOCIETIES MONOGRAPHS COMMITTEE
Harrison W. Craver, Chairman.

PREFACE

In the present volume on "Applied Aero- and Hydromechanics" an attempt has been made to present the more important subjects of the wide field of fluid mechanics in a strictly scientific manner, avoiding a maze of pure mathematical formulas by emphasizing the technical rather than the mathematical treatment. The physical meaning of the various problems has always been brought to the fore, and, whenever possible, experience has been correlated with the fundamental laws and the underlying theories.

As in the case of the "Fundamentals of Aero- and Hydromechanics" (McGraw-Hill, 1934) this volume has been reviewed by Dr. Prandtl and he has added many valuable remarks. It may be mentioned here that Arts. 79 to 81 were written by Dr. Prandtl himself.

The first and the second chapters containing the elements of hydrodynamics and the laws of mechanical similarity are written in rather close agreement with the lectures of Dr. Prandtl. In the third chapter, dealing with the flow through pipes and channels, the author has incorporated considerable material of his own and has represented the subject in quite an extensive manner by making free use of contemporary literature. The fourth chapter on boundary layers again is written in closer agreement with Dr. Prandtl's lectures; whereas the fifth chapter, on the drag of bodies moving through fluids, goes in many respects beyond the discussion given by Dr. Prandtl. This is also true for the sixth chapter dealing with the airfoil theory. The last chapter, in which the author gives a survey of experimental methods and apparatus, is not based on Dr. Prandtl's lectures. However, it was strongly felt that in a treatise on aero- and hydrodynamics at least the more important experimental features should be presented.

The references to contemporary literature by no means claim completeness but are given rather to facilitate a more extensive study of each particular subject. In the "Fundamentals" the number of references was intentionally limited since good books on classical hydrodynamics are available in which complete bibliographies are given. In this respect one may refer to Lamb's "Hydrodynamics."

The author wishes to acknowledge his indebtedness to Dr. J. P. Den Hartog for translating the German edition and for his kindness in undertaking the reading of the proofs.

O. G. TIETJENS.

SWARTHMORE, PA.,
January, 1934.

CONTENTS

CHAPTER VI

INTRODUCTION

The Problem of Flow Resistance.—A completely frictionless fluid, as was discussed in Chap. III, of "Fundamentals," represents only an idealized mental picture of an actual fluid. The results obtained when completely neglecting the internal friction can therefore be considered in the most favorable case only as approximations of actual fluid motions. In general, the agreement between the theoretical and experimental results becomes better when the viscosity becomes smaller.

This statement, however, is true only with one important exception (see Art. 55, "Fundamentals"). Only in the cases where the "boundary layer" formed under the influence of the viscosity remains in contact with the body can an approximation of the actual fluid motion by means of a theory in terms of the ideal frictionless fluid be attempted, whereas in all cases where the boundary layer leaves the body, a theoretical treatment leads to results which do not coincide at all with experiment. And it has to be confessed that the latter case occurs most frequently.

A classical example is the problem of the resistance of a body (for instance, a sphere) moving through a liquid with uniform velocity. The theory on the basis of a frictionless fluid discussed in Art. 68 leads to the paradoxical result that the resistance or drag of such a sphere is zero. The reason for the discrepancy is that in the actual case the boundary layer leaves the sphere, so that the picture of the flow is entirely different from the one examined in the theoretical calculation.

Since the hydrodynamics of the frictionless fluid leads to completely useless results regarding the resistance problem, and a consideration of viscosity in the equations of motion until now has offered unsurmountable mathematical difficulties, there remains only the experimental procedure for determining the laws of drag. For this purpose, extensive series of tests have been carried through, especially for air and water. Such experimenting was greatly accelerated since the beginning of this century by the enormous development in aeronautics, which created great interest in the knowledge of the forces exerted by the air on the airplane or airship.   Because such experiments in general are made by suspending models of airplanes or airships in artificially produced air currents, it became important to know the laws governing the mechanical similarity of the phenomena in the model as compared with the full-size airplane.

Before starting the discussion of the laws of similarity, the elements of the theory of flow and the principles of internal friction of a fluid will be considered briefly, especially for those readers who have not read the "Fundamentals."

APPLIED HYDRO- AND AEROMECHANICS

CHAPTER I

ELEMENTS OF HYDRODYNAMICS

1. The Equation of Euler for One-dimensional Flow.—When dealing with fluid motions one of the most useful conceptions is that of a streamline, which is a curve whose direction in each point coincides with the direction of the velocity of the fluid. In the usual case of continuous velocity distribution, all streamlines passing through a small closed curve form a so-called "stream tube."

A special case of fluid motion is the one in which at any point of space the velocity, pressure, density, etc., remain constant with time. This evidently makes the picture of streamlines also invariable and such a flow is called "steady."

Since the streamlines always have the direction of the velocity, the stream tubes in the steady-state case behave like solid tubes through which the fluid passes. From the law of conservation of matter, it follows that the amount of fluid flowing through each section of such a stream tube per unit of time must be a constant. When A denotes the cross section of the stream tube, ρ the density (which is not necessarily constant), and w the velocity, the so-called equation of continuity for a stream tube becomes (Fig. 1)

Fig. 1.—Stream tube.

$$\rho A w = \text{const.} \tag{1}$$

Now we shall derive an important dynamical relation for the case of a frictionless fluid. To this end we consider an element of the fluid having the shape of an infinitesimally small cylinder (Fig. 2) inside a stream tube. The fundamental law of mechanics, stating that the product of mass and acceleration equals the

1

sum of the forces, is valid for each particle of the fluid. By applying this law to the cylinder of Fig. 2, we obtain

$$\underbrace{\rho dA ds \cdot \frac{Dw}{dt}}_{\text{mass} \times \text{acceleration} =} = \underbrace{\rho g dA ds \cos \alpha}_{\text{gravity force} +} + \underbrace{dA\left\{p - \left(p + \frac{\partial p}{\partial s}ds\right)\right\}}_{\text{pressure force}}.$$

In this formula ρ is the density and g the acceleration of gravity.

The "substantial" or total acceleration Dw/dt in the longitudinal direction of a particle of fluid is generally composed of two terms:

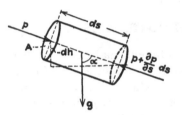

Fig. 2.—Forces on an element of ideal fluid.

1. The change in velocity per second caused by the fact that the velocities at the various points vary with the time, $\partial w/\partial t$. This may be called the "local" differential coefficient.

2. The change in velocity per unit of time caused by the fact that each particle owing to its motion gets into a region where the velocity is different ("differential coefficient of convection"). The expression of the change in velocity with the location is $\partial w/\partial s$ so that the differential coefficient of convection becomes $w\partial w/\partial s$, the change in location per unit of time being represented by the velocity w of the particle.

Therefore, the substantial differential coefficient becomes

$$\frac{Dw}{dt} = \frac{\partial w}{\partial t} + w\frac{\partial w}{\partial s}.^{1}$$

Substituting this expression for the acceleration of a particle in the above equation and dividing by ρdA, we obtain

$$\frac{\partial w}{\partial t}ds + w\frac{\partial w}{\partial s}ds = gds \cos \alpha - \frac{1}{\rho}\frac{\partial p}{\partial s}ds.$$

But for the factor ds this is the equation of Euler for one-dimensional motion.

[1] Mathematically this can be derived also by considering $w = f(t, s)$, and writing for the total differential

$$Dw = \frac{\partial w}{\partial t}dt + \frac{\partial w}{\partial s}ds,$$

$$\frac{Dw}{dt} = \frac{\partial w}{\partial t} + \frac{\partial w}{\partial s} \cdot \frac{ds}{dt} = \frac{\partial w}{\partial t} + w\frac{\partial w}{\partial s}.$$

2. The Equation of Bernoulli for One-dimensional Flow; Three-dimensional Equation of Euler.—Assuming further that, first, the flow is steady, *i.e.*, $\partial w/\partial t = 0$, and, second, the fluid is homogeneous and incompressible, *i.e.*, ρ is constant, we obtain by integration with respect to s,

$$\frac{w^2}{2} + gh + \frac{p}{\rho} = \text{const.} \tag{2a}$$

This integration has been performed along a streamline. Further, we have set $ds \cos \alpha = -dh$ (Fig. 2). The equation (2a), which is of fundamental importance for frictionless fluids, gives the relation between velocity, location, and pressure of those particles of the fluid which are on the same streamline. It is known as the "equation of Bernoulli." For the case that no free surfaces occur and that ρ is constant in the entire fluid, the equation can be simplified somewhat when we denote by p not the absolute pressure but rather the difference between the actual pressure and the pressure which would exist if the fluid were at rest. In that case, the equation of Bernoulli takes the form

$$\frac{w^2}{2} + \frac{p}{\rho} = \text{const.} \tag{2b}$$

It is noted especially that the constant is not necessarily the same for different streamlines.

In the previous discussions, the internal friction or viscosity (which any real fluid possesses to some degree) has been neglected, but even for fluids which have very small viscosity and which can practically be considered as frictionless, there are regions in the field of flow where the friction forces assume such magnitudes that the assumption of no friction is not even approximately true. Such regions occur always in the direct proximity of the bodies along which the fluid flows. There the friction forces assume importance the same as or even greater than the inertia forces (mass times acceleration) which we have considered exclusively until now.

In the case of general three-dimensional fluid motions the equation of Euler is a vector equation, which can be decomposed into three equations for the x-, y-, and z-directions respectively, as shown in Art. 56, "Fundamentals."[1]

[1] TIETJENS, O. G., "Fundamentals of Hydro- and Aeromechanics," Based on Lectures of L. Prandtl, New York, 1934.

$$
\left.
\begin{aligned}
\frac{\partial u}{\partial t} + u\frac{\partial u}{\partial x} + v\frac{\partial u}{\partial y} + w\frac{\partial u}{\partial z} &= X - \frac{1}{\rho}\frac{\partial p}{\partial x}, \\
\frac{\partial v}{\partial t} + u\frac{\partial v}{\partial x} + v\frac{\partial v}{\partial y} + w\frac{\partial v}{\partial z} &= Y - \frac{1}{\rho}\frac{\partial p}{\partial y}, \\
\frac{\partial w}{\partial t} + u\frac{\partial w}{\partial x} + v\frac{\partial w}{\partial y} + w\frac{\partial w}{\partial z} &= Z - \frac{1}{\rho}\frac{\partial p}{\partial z},
\end{aligned}
\right\}
\tag{3}
$$

where u, v, and w are the velocity components in the x-, y-, and z-directions, and X, Y, and Z are the body forces per unit volume in these three directions.

3. Definition of Viscosity; Equation of Navier-Stokes.—In order to obtain a physical picture of the friction in fluids, we consider first the motion of a fluid between two flat parallel plates of which the one moves relatively to the other (Fig. 3).   Assuming the lower plate to be at rest and the upper plate to be moving with a velocity u_1 from left to right, the experiment leads to the following observations: (1) the fluid sticks to the surfaces of the plate; (2) the change of the velocity between the plates is linear (in our case the velocity at any point between the plates is proportional to the distance of this point from the lower plate); (3) the internal friction of the fluid causes a resistance to the motion of the upper plate which is proportional to the gradient of the velocity; there is a force per unit area or a shear stress τ of the magnitude

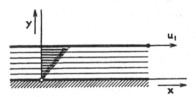

Fig. 3.—Velocity distribution in a viscous fluid between two plates of which one moves relatively to the other.

$$
\tau = \mu\frac{du}{dy},
$$

where μ is a factor of proportionality which indicates the amount of viscosity.   It is a constant for each fluid depending very much upon the temperature and is known by the name of "coefficient of viscosity."   In an elastic medium, the shear stress is proportional to the angular deformation γ:

$$
\tau = G\gamma,
$$

with

$$
\gamma = \frac{\partial \xi}{\partial y},
$$

where G is the modulus of elasticity in shear, and ξ is the displacement of a point in the x-direction. On the other hand, in a fluid medium the shear stress is proportional to the rate of change of angle $\partial\gamma/\partial t$, *i.e.*,

$$\tau = \mu\frac{\partial\gamma}{\partial t} = \mu\frac{\partial\frac{\partial\xi}{\partial y}}{\partial t} = \mu\frac{\partial}{\partial y}\left(\frac{\partial\xi}{\partial t}\right);$$

$$\xi = ut;$$

therefore

$$\frac{\partial\xi}{\partial t} = u.$$

This leads to the relation

$$\tau = \mu\frac{\partial u}{\partial y}, \tag{4}$$

which has been verified by experiment, as will be discussed in Art. 12.

The general differential equations of fluid motion including the effect of viscosity are known as "the equations of Navier-Stokes." A derivation of them can be found in Chap. XV, "Fundamentals."[1] For the x-direction the equation is

$$\frac{\partial u}{\partial t} + u\frac{\partial u}{\partial x} + v\frac{\partial u}{\partial y} + w\frac{\partial u}{\partial z} = X - \frac{1}{\rho}\frac{\partial p}{\partial x} + \frac{\mu}{\rho}\left(\frac{\partial^2 u}{\partial x^2} + \frac{\partial^2 u}{\partial y^2} + \frac{\partial^2 u}{\partial z^2}\right). \tag{5}$$

Two corresponding equations hold for the y- and z-directions. It is seen that for no viscosity ($\mu = 0$), the equation of Navier-Stokes (5) reduces to Euler's equation (3).

[1] See footnote, p. 3.

CHAPTER II

LAWS OF SIMILARITY

4. The Law of Similarity under the Action of Inertia and Viscosity.—In the study of mechanical similarity the question comes up: Under what conditions will a geometrically similar flow of a liquid or gas occur around geometrically similar bodies? For instance, considering the flow of two different fluids (of which one may be a gas) round two spheres of different size, the question is: What conditions have to be fulfilled in order to make the streamline picture in both cases geometrically similar? (Fig. 4.) The answer evidently is that in similar points of the two fields of flow, the forces acting on an element must bear the same ratio to each other at any instant. Depending on the nature of the various forces acting in the fluid, this condition gives us the various laws of mechanical similarity. The first and most important case is that all forces except inertia and frictional forces can be neglected. This case includes the assumption that the liquid or the gas can be considered incompressible; further, that no free surfaces exist and that, hence, the action of gravity is eliminated by statical buoyancy. When it is desired that the flow around the two spheres of Fig. 4 be similar, it is necessary, as stated above, that the ratio between the inertia force and the frictional force at any instant acting on the two corresponding fluid particles be the same.

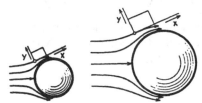

Fig. 4.—Streamlines round two spheres of different size.

Now we proceed to derive the expressions for the inertia force and the frictional force acting on an elementary volume.

An expression for the frictional force per unit volume can be obtained by considering an element of fluid (Fig. 5) whose x-direction coincides with the direction of motion. The difference between the shear forces acting on the element then becomes

6

$$\left(\tau + \frac{\partial \tau}{\partial y}dy\right)dxdz - \tau dxdz = \frac{\partial \tau}{\partial y}dxdydz,$$

or the viscosity force per unit volume is equal to

$$\frac{\partial \tau}{\partial y} = \mu\frac{\partial^2 u}{\partial y^2}.$$

The corresponding expression for the inertia force per unit volume is equal to the product of mass and acceleration per unit volume. When u denotes the velocity component of the fluid particle in the x-direction, the x-component of the acceleration for a steady motion can be represented by the expression $u\frac{\partial u}{\partial x}$ and therefore the inertia force per unit volume becomes $\rho u\frac{\partial u}{\partial x}$ (to be exact, the terms $\rho v\frac{\partial u}{\partial y}$ and $\rho w\frac{\partial u}{\partial z}$ should be added, and for non-steady motions the term $\rho\frac{\partial u}{\partial t}$;

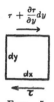

FIG. 5.— Shear stresses on an element.

however, for geometrically similar flows, these terms behave exactly like the term chosen above). The criterion for mechanical similarity therefore is that the ratio of the inertia force and the frictional force

$$\frac{\text{Inertia force}}{\text{Frictional force}} = \frac{\rho u\dfrac{\partial u}{\partial x}}{\mu\dfrac{\partial^2 u}{\partial y^2}},$$

is equal for points similarly situated with respect to the bodies. How do these forces change with a change in the characteristic quantities: the velocity V of the fluid at a great distance from the sphere, the radius a, the density ρ, and the viscosity μ? Evidently the velocity u at any point of the field of flow is proportional to the velocity V of the undisturbed flow (the change from one system of flow to another entails only a change in the unit of time employed). Denoting by $\sim$ proportionality of the two quantities on either side of the sign, we can write

$$u \sim V.$$

For the same reason the differences between velocities at corresponding points are proportional to the velocity V, *i.e.*,

$$du \sim V.$$

The distances between two points in mechanically similar flows are evidently proportional to the dimensions of the bodies about which the flow takes place (for instance, to the radii in the case of spheres).   Therefore the expression $\partial u/\partial x$ is proportional to V/a, and the inertia force itself being represented by $\rho u \dfrac{\partial u}{\partial x}$ is proportional to

$$\rho \frac{V^2}{a}.$$

For the same reasons

$$\frac{\partial^2 u}{\partial y^2} = \frac{\partial}{\partial y}\left(\frac{\partial u}{\partial y}\right) \sim \frac{V}{a^2},$$

so that the frictional force is proportional to

$$\frac{\mu V}{a^2}.$$

The ratio of the inertia force and the frictional force then becomes

$$\frac{\text{Inertia force}}{\text{Frictional force}} = \frac{\rho u \dfrac{\partial u}{\partial x}}{\mu \dfrac{\partial^2 u}{\partial y^2}} = \frac{\rho \dfrac{V^2}{a}}{\mu \dfrac{V}{a^2}} = \frac{\rho}{\mu} V a.$$

Therefore, if in two different flows around geometrically similar bodies the quantity $\dfrac{\rho}{\mu} V a$ is the same, it is to be expected that the streamlines themselves are also geometrically similar. This is the statement of the law of mechanical similarity.   For instance, if we compare two flows of the same fluid of the same temperature and density (μ/ρ = constant) around two spheres of which the one is twice as large as the other, the pattern of the flow is geometrically similar in the two cases if the velocity about the larger sphere is half as great as the velocity about the smaller sphere, because in that case $\dfrac{\rho}{\mu} V a$ has the same value.

Since the quantity $\dfrac{\rho}{\mu} V a$ represents the ratio of two forces, it is a dimensionless number and therefore independent of the units used.   This can be seen immediately by considering the dimensions of the quantities in question.   In the so-called engineering system of units we have

$$[\rho] = \left[\frac{\gamma}{g}\right] = \frac{FT^2}{L^4}, \quad [V] = \frac{L}{T}, \quad [a] = L \text{ and } [\mu] = \frac{FT}{L^2},$$

where L is length, T is time, F is force. It follows that

$$\left[\frac{\rho}{\mu}Va\right] = \frac{FT^2}{L^4} \cdot \frac{L^2}{FT} \cdot \frac{L}{T} \cdot L = 1.$$

Since the quantities ρ and μ often appear as the ratio ρ/μ, this ratio has been given the symbol ν, called the "kinematic viscosity." The dimension of ν therefore is L^2/T. This law of similarity was first found by Osborne Reynolds during his investigation of fluid motions through tubes, which will be discussed in Art. 22. Therefore, the quantity

$$\frac{\rho}{\mu}Va = \frac{Va}{\nu}$$

has been called the "Reynolds' number," usually denoted by R. The fundamental importance of the introduction of this dimensionless quantity for the further development of modern hydrodynamics will be discussed later. For a great number of flow phenomena, the Reynolds' number was the key to finding unknown relations between the experimental results obtained.

5. The Law of Similarity under the Action of Inertia and Gravity.—While in Art. 4 it was assumed that gravity does not act (no free surfaces), now a corresponding law of similarity will be derived considering only inertia and gravity forces and neglecting friction and compressibility. Again it is only necessary to express the fact that for mechanically similar flows at similar points the ratio of the forces acting on these points per unit volume is the same. In other words, the ratio between the inertia force and the gravity force acting on points of similar location with respect to the bodies is the same. The gravity force per unit volume is equal to the weight per unit volume $\gamma = \rho g$ (g being the acceleration of gravity). Therefore the necessary condition for mechanical similarity (neglecting viscosity and compressibility) is

$$\frac{\text{Inertia force}}{\text{Gravity force}} = \frac{\rho u \dfrac{\partial u}{\partial x}}{\rho g} = \text{const.}$$

Since $u\frac{\partial u}{\partial x}$ changes like $\frac{V^2}{a}$, as was seen in Art. 4, the last equation can be written

$$\frac{\text{Inertia force}}{\text{Gravity force}} = \frac{V^2}{ag} = \text{const.}$$

Here V is an entirely arbitrary but characteristic velocity for the flow phenomenon under consideration, and a is an arbitrary characteristic length. This law of similarity was first found by William Froude and is known as Froude's law.[1] The ratio V^2/ag again is a dimensionless number usually denoted by F. This law is extensively used where free surfaces occur, thus calling in the influence of gravity, principally in investigations with ship models. For instance, if the size of the model is one one-hundredth of the size of the ship, Froude's law requires that the velocity of the model be one-tenth of the velocity of the ship, in order to make F a constant. Only then are the pattern of the flow and the shape of the waves similar in the model and in the ship.

In the case of Reynolds' law, considering viscosity and inertia (neglecting gravity), mechanical similarity is possible only when with a small model the model velocity is correspondingly larger. Froude's law, however, requires a decrease in velocity with a decrease in model dimensions. It is evident, therefore, that a simultaneous fulfillment of the two laws of similarity with the same fluid is impossible, *i.e.*, for the same fluid there cannot exist a law of similarity considering inertia forces, frictional forces, and gravity forces all at the same time. Using two fluids of different kinematic viscosity, it is possible to make both laws of similarity valid. Practically, however, this is hardly of importance, since no fluids of sufficiently different ν exist. Denoting by the suffix (1) the actual body and by the suffix (2) the correspondingly small model, it follows from

$$\frac{V_1 a_1}{\nu_1} = \frac{V_2 a_2}{\nu_2}$$

and

$$\frac{V_1^2}{a_1 g} = \frac{V_2^2}{a_2 g}$$

[1] FROUDE, *Trans. Inst. Naval Arch.*, vol. 11, p. 80, 1870.

that the kinematic viscosities of the two fluids must bear the ratio

$$\frac{\nu_1}{\nu_2} = \left(\frac{a_1}{a_2}\right)^{3/2}$$

or

$$\frac{\nu_1}{\nu_2} = \left(\frac{V_1}{V_2}\right)^{3}.$$

In ship-model tests, the resultant of all viscosity forces, *i.e.*, the "skin friction," is in general of the same order of magnitude as the inertia and gravity forces (pressure resistance and wave resistance). According to Froude's procedure, the frictional resistance of the model, as determined by separate tests, is subtracted from the total resistance measured on the model. The rest of the model resistance (the "residuary resistance") is then transformed by means of Froude's rule to the full-size ship. Finally the skin friction of the ship is added to it. This procedure, however, is fairly inaccurate, since even the residuary model resistance mentioned above is not entirely independent of the viscosity, the inaccuracies being greater for smaller models. For this reason, ship builders use relatively large models (15 ft and more).

In both laws of similarity, Reynolds' and Froude's, it was supposed that the effects of compressibility are so small that they can be neglected. In Chap. XIII, "Fundamentals,"[1] it was shown under which conditions gases can be treated as incompressible fluids. In case the effects of compressibility are so large that they are of definite importance (very large velocities or differences in height), it is possible to derive a law of similarity considering only inertia and compressibility. However, in this case also, it appears that the consideration of a third factor (for instance, gravity or viscosity) makes it impossible to fit all conditions.

Since for most meteorological applications a combination of inertia, gravity, and compressibility occurs, it is not possible to study these phenomena by means of model tests.

6. Relation between Considerations of Similarity and Dimensional Analysis.—Since all physical laws can be expressed in a form in which only pure numbers appear which are independent of the units of measurement used, any consideration of similarity can be replaced by a dimensional analysis. Of the quantities

[1] See footnote, p. 3.

appearing in the equation of Navier-Stokes, Eq. (5), the unit of time is determined by the choice of the unit of velocity V and by the unit of length a. On the other hand, the pressure is of no importance for the geometrical similarity of the flow. The only quantities which determine the streamline picture are therefore the velocity V, the length a, the mass per unit volume ρ, and the viscosity μ. We consider the technical system of units with the unit of force F, the unit of length L, and the unit of time T. The question of dimensional analysis is whether there exists a combination

$$V^\alpha a^\beta \rho^\gamma \mu^\delta,$$

which is a pure number. This requires a determination of α, β, γ, and δ such that[1]

$$[V^\alpha a^\beta \rho^\gamma \mu^\delta] = F^0 L^0 T^0 = 1.$$

Since, however, a dimensionless number raised to an arbitrary power remains a pure number, one of the quantities α, β, γ, δ is arbitrary. Putting therefore $\alpha = 1$ we get, substituting for the various physical quantities their dimensions,

$$[V a^\beta \rho^\gamma \mu^\delta] = \frac{L}{T} \frac{L^\beta F^\gamma T^{2\gamma} F^\delta T^\delta}{L^{4\gamma} \quad L^{2\delta}} = F^0 L^0 T^0.$$

Equating the exponents of F, L, and T right and left, equations for β, γ, and δ are obtained, namely,

$$\gamma + \delta = 0.$$
$$1 + \beta - 4\gamma - 2\delta = 0.$$
$$2\gamma + \delta - 1 = 0.$$

This leads to the solution

$$\beta = 1, \qquad \gamma = 1, \qquad \delta = -1,$$

i.e., the only possible dimensionless combination of V, a, ρ, and μ is

$$V a \cdot \frac{\rho}{\mu} = R.$$

If it had been known in advance that ρ and μ only appear in the combination μ/ρ, i.e., $\delta = -\gamma$, the derivation would have been still simpler. Since, therefore,

---

[1] A quantity in square brackets means the "dimension" of that quantity.

$$\left[\frac{\mu}{\rho}\right] = [\nu] = \frac{L^2}{T}$$

and

$$[Va] = \frac{L^2}{T},$$

it follows that Va/ν is the only possible combination giving a pure number.

Although this dimensional analysis physically is not so instructive as the similarity consideration, it has the advantage of being still applicable when the exact equation of motion is unknown and we know only which physical quantities are of importance for the phenomenon.

CHAPTER III

FLOW IN PIPES AND CHANNELS

A. LAMINAR FLOW

8. General.—Investigations of the flow phenomena in pipes and channels were performed early; in fact this is the most important subject of hydraulics. Since the laws of internal friction of fluids were unknown, it was necessary to be contented with experimental results pertaining to each individual case. These individual experiments, however, could not be coordinated into a law.

In the middle of the nineteenth century, an exact solution of the hydrodynamical equations was found for the flow of a fluid through straight tubes of circular cross section, taking into account the influence of viscosity. This is one of the very few cases in which a complete integration of the general differential equations of viscous fluids has been accomplished. However, it appeared that this solution hardly solved the difficulties of practical hydraulics any better, since the conditions under which it is valid do not occur often. In the large majority of cases of flow through pipes or channels, especially in technical applications, the solution found does not apply. This is due to the fact that there exist two radically different kinds of flow. Considering, for instance, the flow through a glass tube, using water in which small particles are suspended, it is seen that most often the particles of fluid do not move in paths parallel to the walls of the tube but flow through in a very irregular manner. Besides the principal motion in the direction of the axis of the tube, secondary motions perpendicular to the axis can be observed. This kind of flow is called "turbulent flow." The majority of cases of fluid flow including those occurring in technical applications are of this kind. When in our experiment with the glass tube the flow of water is throttled down more and more, there will be a certain small velocity at which the individual particles of fluid start moving regularly in paths parallel to the walls of the tube. This is the second kind of flow referred to. It is commonly called "laminar flow," and to it applies the solution of the hydrodynamical equations mentioned before.

9. The Fundamental Investigation of Hagen.—Though the existence of the two forms of flow, turbulent and laminar, was known for a long time, the first systematic tests to discover the laws of these two phenomena were not made until the middle of the last century. Very accurate experiments were carried out by G. Hagen, who deserves great credit for his work though it did not become widely known. This is probably due to the fact that he published his results in terms of "Prussian ounces," "Parisian inches," etc., which require a considerable amount of calculation before they can be compared with the results of more modern investigators. The first of his two publications (1839) is limited to laminar flow only.[1] Hagen used for his tests three brass tubes of various diameters[2] and expressed the measured pressure heads h of his supply tank as a function of the weight of the water flowing out per second W. He made the assumption

$$h = h_1 + h_2 = aW + bW^2$$

and showed that a and b are constants for each tube; a was found to be very much dependent on the temperature, while b is independent of it. Showing a good understanding of the physical phenomenon, Hagen observed that the part $h_2 = bW^2$ of the total head is used for imparting kinetic energy to the fluid while the part $h_1 = aW$ is necessary for overcoming the friction resistance.

Therefore, when only friction comes into consideration, the pressure head is proportional to the rate of flow where the factor of proportionality depends very much on the temperature. Using the method of least squares, the relation between the quantity a and the temperature was determined from the experimental data and the value a for the various tubes reduced to a definite temperature (10°C). Dividing the expression for h by the lengths of the tubes, the factors of proportionality a and b thus transformed were found to be inversely proportional to the fourth power of the tube diameter. When r denotes the radius

[1] HAGEN, G., On the Motion of Water in Narrow Cylindrical Tubes (German), *Pogg. Ann.*, vol. 46, p. 423, 1839.

[2] Diameters, 0.255 cm, 0.401 cm, 0.591 cm; lengths, 47.4 cm, 109 cm, 105 cm., respectively.

of the tube, he found (in terms of Parisian inch, Prussian ounce seconds) that

$$h = h_1 + h_2 = 0.000009117\frac{lW}{r^4} + 0.0002056\frac{W^2}{r^4}.$$

Considering only the term proportional to the first power of W, which is taken up by friction, it is seen that the weight of water delivered per second is proportional to the pressure head h_1 and to the fourth power of the radius, and inversely proportional to the length of the tube. Introducing into the above relation the mean velocity $\bar{u}$ instead of the weight delivered per second ($W = \pi r^2 \bar{u}\gamma$), and, instead of the pressure head h, the pressure difference $\Delta p = h\gamma = h\rho g$, we get (in c g s units)

$$\Delta p = \Delta p_1 + \Delta p_2 = 0.103\frac{l\bar{u}}{r^2} + 1.35\rho\bar{u}^2.$$

The numerical factors in this result have been obtained with one Parisian inch equal to 2.707 cm and the specific gravity of water as 1.355 Prussian ounces per cubic Parisian inch. Introducing in place of the coefficient 0.103, the viscosity μ, which for the temperature 10°C is[1] 0.013 g/cm sec., the above formula transforms to

$$\Delta p = \Delta p_1 + \Delta p_2 = 8\mu\frac{l\bar{u}}{r^2} + 2.7\frac{\rho\bar{u}^2}{2}. \tag{1}$$

For the relation between the viscosity and the temperature (between 0° and 20°C), Hagen gives

$$\mu = 0.01800 - 0.000655t + 0.0000144t^2$$

expressed in c g s units and in degrees centigrade. In Fig. 6 some of the values calculated by means of this formula are compared with the best and most up-to-date results of Thorpe and Rodger, as well as with those of Bingham and White,[2] showing the remarkable accuracy of Hagen's tests.

10. The Investigation of Poiseuille.—Approximately simultaneous with Hagen's publication in *Poggendorfs Annalen*, the Parisian physician and physicist Poiseuille experimentally found the same law for the laminar flow of water through very narrow

---

[1] THORPE and RODGER, *Phil. Trans. Roy. Soc.*, vol. 185, Plate 8, 1894.

[2] BINGHAM and WHITE, *Z. physik. Chem.* (German), vol. 80, p. 670, 1912.

capillar tubes of glass. With the object of studying the movement of the blood through capillary veins Poiseuille investigated the rate of flow as influenced by the pressure drop, the length of the capillary, its diameter, and the temperature of the fluid. In three preliminary publications in 1840 and 1841 (*i.e.*, two years after Hagen), as well as more detailed in 1846, Poiseuille derived from his very careful experiments the law that the rate of flow is proportional to the pressure drop, to the fourth power of the radius, and inversely proportional to the length of the tube. With this result, Poiseuille had shown that the law found by Hagen for laminar motion in tubes is also valid for capillary tubes. How-

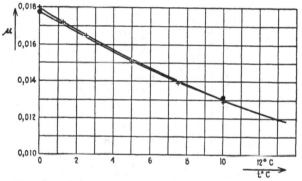

Fig. 6.—Viscosity of water as a function of the temperature. The upper curve after Bingham and White; the lower one after Thorpe and Rodger. The + points by Hagen (1839); the ⊗ by Poiseuille (1841).

ever, the knowledge that a part of the pressure is used to impart kinetic energy to the fluid was lacking with Poiseuille, whereas Hagen expressed this thought very clearly. Poiseuille observed only that his law ceases to be valid when the length of the tube is less than a certain multiple of the diameter. For instance, he published the statement that for a capillary of 0.29-mm diameter (about three times the dimension of a capillary vein) his law ceases to hold for a length shorter than 2 mm. For such short tubes (measured in diameters) the pressure head necessary for overcoming internal friction becomes so small that it is not possible to neglect the second part of the pressure head necessary for creating kinetic energy.

11. The Law of Hagen-Poiseuille.—In view of the fact that Hagen published the law of laminar flow two years in advance of Poiseuille and moreover that Hagen calculated from his experi-

ments a correction term for the kinetic energy, this law is called
the law of Hagen-Poiseuille since it was derived by both independently.

Neglecting the correction term for kinetic energy, the experiments on laminar flow have led to

$$\Delta p = 8\mu \frac{l}{r^2}\bar{u}.$$

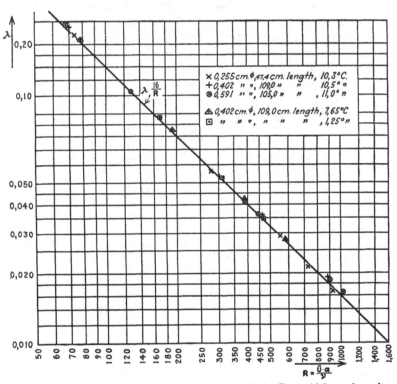

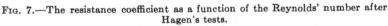

Fig. 7.—The resistance coefficient as a function of the Reynolds' number after
Hagen's tests.

The pressure drop in tubes with turbulent flow had been investi-
gated even earlier and was found to be proportional to $\frac{l}{r} \cdot \rho\frac{\bar{u}^2}{2}$.
Though for laminar flow the pressure is proportional only to the
first power of the velocity, the proportionality with $\frac{l}{r} \cdot \rho\frac{\bar{u}^2}{2}$ of the
turbulent flow has been applied also to the laminar flow.   With

this procedure, the proportionality factor λ naturally cannot be constant any more. We have

$$\Delta p = \lambda \cdot \frac{l}{r} \cdot \rho \frac{\bar{u}^2}{2} \qquad (2)$$

and, using the above formula for Δp,

$$\lambda = \frac{16}{r\bar{u}} \cdot \frac{\mu}{\rho} = \frac{16}{\underbrace{\frac{r\bar{u}}{\nu}}} = \frac{16}{R}, \qquad (2a)$$

where R again is the Reynolds' number.

Plotting λ as a function of R on double logarithmic paper, the relation gives a straight line of 45-deg. slope passing through $\lambda = 0.16$, for $R = 100$. Logarithmic paper is used in order to prevent too great a crowding of the smaller Reynolds' numbers.

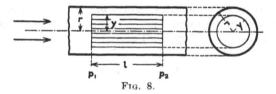

FIG. 8.

In Fig. 7 the values of λ calculated from Hagen's experiments corrected for kinetic energy have been plotted against R. It is seen that the various values conform very well to the straight line $\lambda = 16/R$, although the measurements have been carried out with tubes of greatly differing diameters and lengths and over a wide range of temperatures.

12. Derivation of Hagen-Poiseuille's Law from Newton's Viscosity Law.—In order to derive the law of Hagen-Poiseuille from Newton's differential expression for viscosity (page 4), we shall consider a cylindrical piece of fluid inside the tube (Fig. 8). The pressure difference $p_1 - p_2$ between the two faces of the cylinder causes a longitudinal force excess $(p_1 - p_2)\pi y^2$, which leads to a certain shear stress τ on the curved surface of the cylinder. For the case of steady, non-accelerated flow, we have

$$(p_1 - p_2)\pi y^2 = 2\pi y l \tau$$

or

$$\tau = \frac{p_1 - p_2}{l} \cdot \frac{y}{2}.$$

Considering that du/dy is negative, and substituting Newton's friction law $\tau = \mu \dfrac{du}{dy}$, we have

$$\frac{du}{dy} = -\frac{p_1 - p_2}{\mu l} \cdot \frac{y}{2}$$

or

$$\int_r^y du = \frac{p_1 - p_2}{2\mu l} \int_y^r y\, dy.$$

Considering further that the fluid adheres to the sides of the tube, i.e., $u(r) = 0$, it follows that

$$u(y) = \frac{p_1 - p_2}{4\mu l}(r^2 - y^2).$$

Therefore, it is seen that in the case of laminar flow through cylindrical tubes of circular cross section the velocity distribution has the shape of a paraboloid of rotation. The maximum velocity for $y = 0$ will be denoted by u_0. Since the volume Q of such a paraboloid equals $\pi r^2 u_0/2$, we have

$$Q = \frac{p_1 - p_2}{8\mu l}\pi r^4$$

or

$$p_1 - p_2 = \Delta p = 8\mu \frac{l}{r^2} \cdot \frac{Q}{\pi r^2}.$$

Introducing finally the mean velocity $\bar{u} = Q/\pi r^2$ we obtain

$$\Delta p = 8\mu \frac{l\bar{u}}{r^2}, \tag{3}$$

which coincides completely with the viscosity term Δp_1 of Eq. (1) on page 16.

The coincidence of the tests for tubes of various diameters with the theoretical Eq. (3) can be considered as an experimental verification of Newton's friction law, stating that the shear stress is proportional to the rate of deformation and also that the fluid sticks to the walls and thus does not flow past them with a finite velocity. Since these experiments can be carried out with great accuracy they are well suited for an experimental determination of the viscosity μ.[1] However, in greatly rarefied gases, where the molecular free path cannot be neglected with respect to the

[1] ERK, S., Viscosity Measurements on Fluids and Investigations on Viscosimeters (German), *Forschungsarbeiten V. D. I.*, vol. 288, 1927.

radius of the tube, discrepancies are found which can be interpreted as slipping along the walls. This fact is in accordance with the theory.

13. Limits of the Validity of the Hagen-Poiseuille Law.—Recently experiments have been made to test the validity of the law for fluids of extremely high viscosity as well as for fluids under very high pressures. The results of Reiger, Ladenburg, and Glaser[1] show that even for fluids of $\mu = 10^6$ (rosin in turpentine) the law holds with great accuracy. However, Glaser's experiments indicate that the law does not hold when the radius of the tube is smaller than a certain limit depending on the viscosity. This limit in c g s units was found to be

μ	r
10^5	0.1
10^7	0.5
10^9	1.0

For radii smaller than these, an important increase in the value for μ was found.

Recently colloidal fluids have also been investigated as to their behavior with respect to the law of Hagen-Poiseuille.[2]

14. Phenomena Near the Entrance of the Tube.—Equation (3), Art. 12, applies to the laminar flow in a tube at a sufficient distance from its entrance. At the entrance itself, which for simplicity's sake we assume to be rounded as shown in Fig. 9, it is clear that no parabolic velocity distribution can exist. The fluid rather enters the tube with a velocity which is constant across the section, and only directly at the wall is the velocity zero;

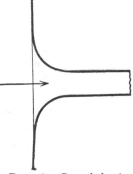

FIG. 9.—Rounded pipe entrance for avoiding entrance eddies.

[1] REIGER, R., On the Validity of Poiseuille's Law for Fluids of High Viscosity and for Solids (German), *Ann. Physik*, vol. 19, p. 985, 1906.

LADENBURG, R., On the Internal Friction of Viscous Fluids and Its Relation with the Pressure (German), *Ann. Physik*, vol. 22, p. 287, 1907.

GLASER, H., On the Internal Friction of Viscous and Plastic Bodies and the Validity of Poiseuille's Law (German), *Ann. Physik*, vol. 22, p. 694, 1907.

[2] REINER, M., The Hydrodynamics of Colloids (German), *Z. angew. Math. Mech.*, vol. 10, p. 400, 1930.

hence a very sudden increase in the velocity from zero takes place in an extremely thin layer near the wall.   Due to the influence of the internal friction, the layers of fluid lying farther away from the surface are retarded, *i.e.*, the boundary layer which was very thin at the entrance of the tube becomes thicker and thicker at larger distances from the entrance.

On the other hand, the volume transported remains the same for each section so that due to the fact that the layers near the wall are retarded, the inner parts near the center of the tube must be accelerated until finally the equilibrium relation between pressure drop and friction resistance (as discussed in the previous article) has adjusted itself.   As was seen before, the theory gives for this condition a parabolic velocity distribution and since a paraboloid of rotation of the same volume as a cylinder on the same base has twice its height, it follows that the velocity in the middle u_0 is accelerated to double the value of the mean velocity $\bar{u}$. Expressing the velocity u by means of the ratio $u/\bar{u}$, it can be stated that at the entrance of the tube, *i.e.*, for $x = 0$, the condition is expressed by $u/\bar{u} = 1$, while with increasing x, the parabolic distribution $\dfrac{u}{\bar{u}} = 2\left[1 - \left(\dfrac{y}{r}\right)^2 \right]$ is approached asymptotically.

Although this distribution is theoretically never reached, it is of interest to know for which value of x the actual velocity distribution differs so little from the parabolic distribution that the velocities in the middle are·not more than 1 per cent apart.   This length of tube will be called the "length of transition."

15. The Length of Transition.—Boussinesq[1] was the first to make a theoretical investigation of these phenomena.   His calculated results are in good agreement with the experiment for velocity distributions at some distance away¯from the entrance of the tube.   However, for sections near to that entrance, his calculated distributions do not check with the experimental ones.   For the length of transition he also finds a value which is in good agreement with the experiments, namely,

$$\frac{x_1}{rR} = 0.26, \tag{4}$$

where x_1 is the length of transition and $R = (\bar{u}.r)/\nu$ is the Reynolds' number.   For instance, in a tube of 0.5-in. diameter, a

[1] BOUSSINESQ, J., *Compt. rend.*, vol. 113, pp. 9 and 49, 1891.

Reynolds' number of about $R = 4,000$[1] and water of 20°C, the above equation requires a length of at least 22 ft in order to have a difference of less than 1 per cent between the actual velocity in the middle and the theoretical one according to Hagen-Poiseuille. Therefore Hagen-Poiseuille's law as expressed by Eq. (3) holds only in sections farther than 22 ft away from the entrance to the tube.

16. The Pressure Distribution in the Region Near the Entrance. In sections of the tube near the entrance, it is necessary to have a larger pressure drop per unit length than is required by Eq. (3), since a part of this drop is utilized for accelerating the inside layers and consequently for increasing the kinetic energy of the flow. Disregard of this fact has often been the reason that experimental results were understood incompletely or found to be

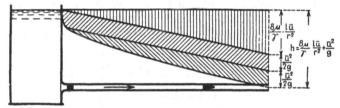

Fig. 10.—Pressure diagram in a pipe flow with constant head in tank.

in disagreement with those of others. In order to explain these phenomena better, we shall consider in Fig. 10 the flow from a large reservoir through a tube with a well-rounded entrance. The reservoir is assumed to be so large that the velocities inside it can be neglected. Denoting by $p_0 = h\gamma$ the pressure in the reservoir at the elevation of the center line of the tube, the pressure at the entrance $x = 0$ will be $p_1 = p_0 - \left(\dfrac{\rho\bar{u}^2}{2}\right)$. The equivalent of this loss in pressure energy is found in the gain in kinetic energy of the flow in the tube. As we know, the practically constant velocity distribution at $x = 0$ is gradually transformed to a parabolic distribution in the region of transition. This, however, is equivalent to a further increase in the kinetic energy to the amount $(\rho\bar{u}^2/2)$ (the flow of kinetic energy through the section πr^2, being $\displaystyle\int_0^r \frac{\rho u^3}{2} \cdot 2\pi y\, dy$, is twice as large for the para-

---

[1] As will be seen in Art. 24, it is possible to obtain a regular laminar flow with Reynolds' numbers of this magnitude when the entrance to the tube is rounded off well.

bolic distribution as for a constant distribution). The total pressure drop used for creating the kinetic energy of the parabolic velocity distribution therefore is $p_0 - p_2 = 2\dfrac{\rho\bar{u}^2}{2}$.

To this has to be added the pressure drop for overcoming the friction in the tube, determined by the formula of Hagen-Poiseuille [Eq. (3)], so that finally

$$\Delta p = p_0 - p_2 = 8\mu\frac{l\bar{u}}{r^2} + 2\frac{\rho\bar{u}^2}{2}, \tag{5}$$

in which p_2 is the pressure in a section with parabolic distribution.

17. The Correction Term for Kinetic Energy.—The term $2\dfrac{\rho\bar{u}^2}{2}$ is often referred to as the correction of Hagenbach, which, however, is not entirely justified. In the first place, Neumann[1] gave the complete Eq. (5) in his lectures (published by Jacobson[2] prior to Hagenbach's paper[3]). Further there is an error in Hagenbach's publication,[3] owing to which he does not obtain the correction term of Eq. (5) but rather $2^{3/5}\dfrac{\rho\bar{u}^2}{2}$, which is too small. From his experimental investigations Hagen had already recognized the importance of a correction term for the kinetic energy and had found it to be $2.7\dfrac{\rho\bar{u}^2}{2}$. This value is considerably too large, which is due to the fact that Hagen did not use a rounded entrance but one which was squarely cut off. This caused a contraction of the jet, with subsequent spreading out again, leading to an additional pressure drop.

In Eq. (5), it was assumed that Hagen-Poiseuille's law is valid in the region near the entrance in spite of the fact that the velocity distribution in this part is considerably different from the theoretical parabola. A justification, however, for this assumption cannot be given. It is rather probable that the pressure drop for overcoming the friction in the entrance region is larger than the corresponding pressure drop for the final parabolic

[1] NEUMANN, F., "Introduction to Theoretical Physics" (German); lectures given in 1859–1860, Leipzig, 1883.

[2] JACOBSON, H., Contributions to Haemodynamics (German), *Arch. Anat. Physiol.*, p. 80, 1860.

[3] HAGENBACH, E., On the Determination of the Viscosity of a Fluid by Flow Experiments through Tubes (German), *Pogg. Ann.*, vol. 109, p. 385, 1860.

distribution. However, the accuracy of the experiments made so far is not yet sufficient to decide this question.

18. The Velocity Distribution in the Region Near the Entrance. In order to make a theoretical investigation of the phenomena near the entrance of the tube, L. Prandtl suggested studying the equilibrium equation between the change in momentum, the pressure drop, and the friction force acting on an elemental slice perpendicular to the direction of flow, assuming a certain velocity distribution. The assumption made for this distribution consisted of a constant middle part bounded by two parabolic arcs, as shown in Fig. 11. At the entrance of the tube, the width of the parabolic arcs was zero. This width increased with the distance from the entrance until, at a certain point, the arcs were united into a single parabola. The constant velocity in the middle had to increase at such a rate that the same volume of water was flowing through all sections. The constant velocity core was made to satisfy the equation of Bernoulli, while the momentum theorem was satisfied for the total cross section. The calculation, as carried out by L. Schiller,[1] gave an excellent agreement of the various velocity-distribution curves with subsequent experimental measurements, at least for the first third of the

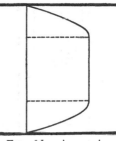

Fɪɢ. 11.—Approximation of laminar velocity distribution by a straight line and two parabolic arcs.

length of the transition region, which is its most important part.

At larger distances from the entrance, the velocity in the core increases slower than indicated by the calculation of Schiller. Moreover, measurements have shown that the flow in the core is constant in cross sections only near to the entrance (where the boundary layer has not yet become too thick), while for sections farther away from it, the flow in the core has first a slight and later a more pronounced curvature. Figure 12 shows the development of the laminar velocity distribution for a rounded entrance, according to experiments by J. Nikuradse. It is seen that until (about) $x/rR = 0.04$, the assumption of a central flow independent of the friction and of a parabolic drop in velocity

[1] Sᴄʜɪʟʟᴇʀ, L., Investigations on Laminar and Turbulent Flow (German), *Forschungarbeiten V. D. I.*, vol. 248, 1922; *Z. angew. Math. Mech.* (German), vol 2, p. 96, 1922; or *Physik. Z.* (German), vol. 23, p. 14, 1922.

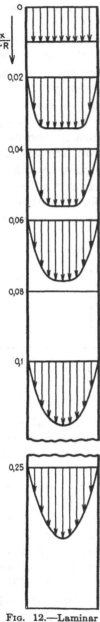

FIG. 12.—Laminar velocity distribution near entrance of pipe after tests by Nikuradse.

toward the walls is justified; however, from there on a definite core flow, where the friction has not made itself felt yet, does not exist. Figure 13 shows a number of curves for which the dimensionless velocity $u/\bar{u}$ is plotted against x/rR, for various distances y/r from the axis. For $y/r = 0$, *i.e.*, for the velocity in the axis of the tube, the theoretical values obtained by Boussinesq and Schiller are drawn in as dotted lines. It is seen that Schiller's curve agrees fairly well with the experiments until (about) $x/rR = 0.05$; however, his result for the length of the transition region $x/rR = 0.115$ is considerably too small. On the other hand, the values of Boussinesq do not check near the entrance but give better agreement from $x/rR = 0.1$ on, where the velocity curves show a more parabolic shape. Also Boussinesq's value for the length of the transition region $x/rR = 0.26$ seems to be in agreement with the available experimental results.

19. The Pressure Drop in the Entrance Region in the Case of Laminar Flow.—For the kinetic-energy correction in the total pressure drop in the region of transition at the entrance, the theories of both Schiller and Boussinesq give values that are too large. Schiller gives the value $2.16\rho\dfrac{\bar{u}^2}{2}$ while Boussinesq finds $2.24\rho\dfrac{\bar{u}^2}{2}$. Since both theories are approximate, it is for the experiment to decide the correct value. A very good agreement was obtained between the calculated pressure drop of Schiller and experiments made by himself. In one of those a tube was used of 2.399-cm inside diameter; the first pressure measurement p_1 was made at a distance of 104.15 cm and the second, p_2, at a distance of 196.77 cm from the rounded entrance. Figure 14 shows the theoretical straight line of Hagen-Poiseuille and also the curve passing through the experimental points, showing that in the region near the entrance there are important discrepancies with the Hagen-Poiseuille law. The

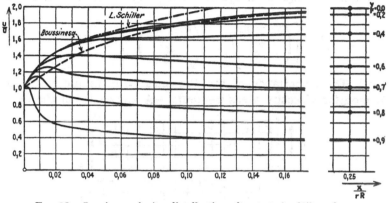

FIG. 13.—Laminar velocity distribution after tests by Nikuradse.

calculated values of Schiller shown in the figure by small open circles are in very good agreement with the experiments. Denoting by p_0 the pressure in the reservoir and by p the pressure at a

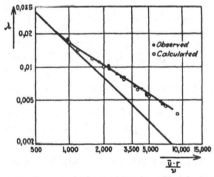

FIG. 14.—Pressure-drop coefficient of laminar flow in entrance region.

distance x from the rounded mouth, Schiller's calculation gives $\dfrac{(p_0 - p)}{\dfrac{\rho \bar{u}^2}{2}}$ as a function of $\dfrac{x}{r\bar{R}}$, shown graphically in Fig. 15.

If the dimensionless pressure drop per unit length λ be defined by

$$\lambda = \frac{p_1 - p_2}{\rho \frac{\bar{u}^2}{2}} \cdot \frac{r}{x_2 - x_1},$$

this quantity can be determined from Fig. 15 by means of the relation

$$\lambda = \left[\left(\frac{p_0 - p_2}{\rho \frac{\bar{u}^2}{2}} \right)_{x_2} - \left(\frac{p_0 - p_1}{\rho \frac{\bar{u}^2}{2}} \right)_{x_1} \right] \cdot \frac{r}{x_2 - x_1}.$$

20. The Importance of the Pressure Drop in the Entrance Region for Viscosity Measurements.—A knowledge of the pres-

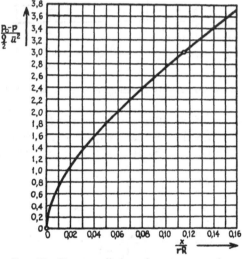

Fig. 15.—Pressure diagram in entrance region.

sure drop in the entrance region is especially important for viscosity determinations by means of the usual Saybolt method. Although in most cases relatively short tubes (small x/rR) are employed, the validity of Hagen-Poiseuille's law is assumed. In case the tubes used are so long that at the end the parabolic velocity distribution is nearly reached, it is generally sufficient to apply the correction for kinetic energy in one of the forms discussed. However, if the tube is so short that x/rR is smaller than about 0.1, Schiller has shown that from the measured pressure drop $p_0 - p_1$ the corresponding

$$\frac{x}{rR} = \frac{x\mu}{r^2\bar{u}\rho}$$

can be found from Fig. 15. With this chart the value of μ can be calculated, as soon as experimental determinations have been made of the volume flowing through per second, the length of the tube, its radius, and the density of the fluid.

B. THE TRANSITION BETWEEN LAMINAR AND TURBULENT FLOW

21. The First Investigations by Hagen.—Already in his first publication on the flow of water through cylindrical tubes (1839) Hagen called attention to the fact that the mode of flow discussed by him ceased to exist when the velocity increased beyond a

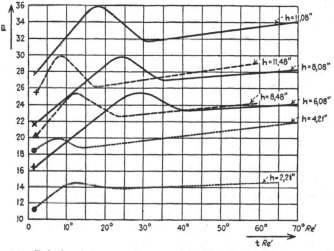

FIG. 16.—Relation between $\bar{u}$ (expressed in Rhineland inch per second) and the temperature (expressed in degrees Réaumur) for various pipe diameters and heads h (in Rhineland inch) after tests by G. Hagen. ——— 0.281 cm. diameter; ——— 0.405 cm diameter; ..... 0.596 cm diameter.

certain limit. He observed that the outflowing jet below this velocity looked like a solid bar of glass; above it the jet commenced to oscillate and the flow ceased to be uniform but came in spurts.

In 1854 Hagen published a second article[1] in which he showed that the transition of the laminar into the turbulent state does not only depend on the velocity but also on the viscosity of the fluid. Using the same tubes as with his first tests, he determined the relation between the volume delivered and the temperature with a constant pressure head for each series of experiments. Figure 16 shows the results of Hagen's tests as given by himself.

[1] HAGEN, G., On the Influence of Temperature on the Movement of Water through Pipes (German), *Abhandl. Akad. Wiss.*, p. 17, Berlin, 1854.

Each curve corresponds to a definite pressure head.   The curves
for the narrow tube (diameter 2.8 mm) are drawn in full; those
for the medium tube (diameter 4 mm) are broken, while for the
wide tube (diameter 6 mm) dotted lines are shown.   In the
upper curve it is seen, for instance, that for a constant pressure
head the mean velocity first increases with increasing temperature
(*i.e.*, decreasing viscosity), then decreases again, reaches a minimum, and increases for the second time.   The first increasing
branch of the curve corresponds to the laminar flow.   The part
of the curve between the maximum and the minimum corresponds to the transition between laminar and turbulent flow,
while the last slowly increasing branch corresponds to the turbulent mode of flow.   Hagen ascertained this by adding sawdust
to the water and observing that for small velocities the wood
particles moved in straight lines through the tube, while for
larger velocities they were thrown from one side to another and
moved quite irregularly.   He considered this irregular motion
to be caused by the irregularities of the tube wall or possibly by
the entrance of the water through the squarely cut end of the
pipe.

In a third publication,[1] Hagen observed several times that the
transition from the turbulent flow to the laminar one depends
on the radius of the tube, on the velocity, and on the temperature
of the water.   The turbulent flow will become laminar as soon
as any one of the three quantities mentioned, or all three of them
together, decrease below a certain limit.

22. The Fundamental Investigation by Reynolds.—Considering the early date of his investigations, Hagen had a very good
conception of the phenomena of laminar and turbulent flow.
However he did not succeed in finding a unifying principle for
plotting the results of his experiments shown in Fig. 16.   The
credit for having found such a principle belongs to Osborne Reynolds.[2]   In his paper of 1883 he showed by means of dimensional
analysis that the transition between laminar and turbulent flow
can depend only on the dimensionless expression

$$\frac{\bar{u}r}{\nu}.$$

[1] Hagen, G., *Abhandl. Akad. Wiss.* (German), Berlin, 1869.

[2] Reynolds, Osborne, An Experimental Investigation of the Circum-
stances Which Determine whether the Motion of Water Will Be Direct or
Sinuous, and of the Law of Resistance in Parallel Channels, *Phil. Trans.
Roy. Soc. London*, 1883, or *Sci. Papers* vol. 2, p. 51.

The law of similarity which later was named after him (Art. 4) expresses the fact that two different motions taking place in two geometrically similar vessels are also mechanically similar when they have the same value of $\bar{u}r/\nu$, the Reynolds' number.[1]

The great simplification obtained in plotting the test results by means of this Reynolds' number can be seen in Fig. 17, which

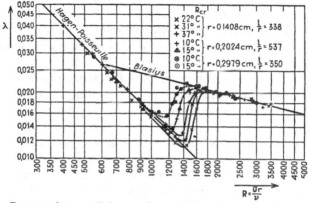

FIG. 17.—Pressure-drop coefficient *vs.* Reynolds' number, being Hagen's tests of Fig. 16 replotted (squarely cut off entrance).

represents the curves of Fig. 16, now plotted on the Reynolds' number as abscissa. The ordinates are also represented by a dimensionless quantity, namely,

$$\lambda = \frac{\Delta p}{\frac{\rho \bar{u}^2}{2}} \cdot \frac{r}{l} = \frac{hg}{\frac{\bar{u}^2}{2}} \cdot \frac{r}{l}.$$

In the plotting from Fig. 16 to Fig. 17, the amounts $2.7\dfrac{\bar{u}^2}{2g}$ for the laminar flow and $1.4\dfrac{\bar{u}^2}{2g}$ for turbulent flow have been subtracted from the ordinates. This is to take care of the correction for kinetic energy, as discussed on page 24. It is seen in Fig. 17 that the replotted curves of Fig. 16 show all the same characteristics: For Reynolds' numbers below about 1,100 to 1,400, the pressure-drop coefficients λ lie on the straight line of Hagen-Poiseuille for laminar motion (Fig. 7). After this, there

[1] This is true for motions where only inertia forces and viscosity forces play a part. It is not true for motions where gravity has to be taken into account, as for instance when free surfaces occur.

is a comparatively sudden increase in the values of λ until a certain maximum is reached. From there on, λ diminishes again with increasing Reynolds' number, however, slower than at first. The region of increasing λ corresponds to the transition between laminar and turbulent flow. For larger Reynolds' numbers the flow is purely turbulent. It is seen that the experimental results lie on a straight line with an inclination of $1:4$ so that, owing to the logarithmic coordinate system used, it follows that the coefficient λ of turbulent flow is proportional to the fourth root of the Reynolds' number. In Art. 30 this relation will be discussed in detail.

Figure 17 shows, moreover, that for a determination of the so-called "critical Reynolds' number" (*i.e.*, the Reynolds' number where the transition between laminar and turbulent flow occurs) pressure-drop measurements are most appropriate. This method was used by Reynolds. Unlike Hagen, who measured the pressure drop between the reservoir and the end of the tube, Reynolds measured the pressure drop in a certain length of tube, having a long stretch between the tank and the location of his measurements. He used tubes of $\frac{1}{4}$ and $\frac{1}{2}$ in. diameter, both about 16 ft long. In either tube the length between the reservoir and the measuring spot was 11 ft, *i.e.*, 528 and 264 diameters respectively.

23. The Critical Reynolds' Number.—From experiments conducted on these two tubes, Reynolds found a complete verification of his theorem that the transition between laminar and turbulent flow takes place at a definite value of $\bar{u}r/\nu$ even for tubes of different diameters. Expressing his test results in terms of the dimensionless quantities λ and $\bar{u}r/\nu,$[1] it is seen that the first deviation of the pressure-drop coefficient λ from the Hagen-Poiseuille law takes place at $\bar{u}r/\nu = 1000-1100$ approximately.[2]

The value of R where turbulence just starts is known as the "critical Reynolds' number" and the mean velocity $\bar{u}$ corresponding to this number is known as the "critical velocity."

What conclusion can now be drawn from the experimental results of Hagen (Fig. 17) and of Reynolds? The common interpretation is that laminar flow or even flow with parabolic

---

[1] Reynolds himself gave his results as log Δp in terms of log $\bar{u}$.

[2] It is noted in passing that the pressure drop per unit length in the turbulent region measured by Reynolds is appreciably smaller than that found by other more recent investigators.

velocity distribution passes into turbulent flow at a Reynolds' number of about 1,000. This statement, however, is incorrect for the following reason: In the first part of the tube adjacent to the reservoir, laminar flow, *i.e.*, flow parallel to the sides of the tube, does not occur. A flow with a parabolic velocity distribution is even less probable since it was shown in Art. 15 that for its generation a fairly long piece of tube is required. Actually the fluid gets into the tube with some initial turbulence, which in Hagen's experiments is due to the sharp entrance and in Reynolds' experiments due to the fact that he connected his test tube to the water faucet. Therefore the actual conditions are such that below the critical Reynolds' number these initial disturbances are damped out, while for larger Reynolds' numbers they develop into the irregular motions which are typical of turbulent flow.

24. Influence of the Initial Disturbance on the Critical Reynolds' Number.—The question presents itself whether this critical number is the same for all tubes and for all experimental set-ups. Some doubt as to this comes up in comparing the results of Hagen with those of Reynolds. Recent experiments, especially those of Schiller, have shown that very small disturbances can be damped out even for values of R, which are appreciably greater. Quite high critical Reynolds' numbers can be obtained by letting the water flowing out of the faucet first come to rest in a large tank and then using a well-rounded entrance to the tube so that no contraction of the jet takes place. In such cases, it can be stated that the laminar flow which is formed close to the entrance of the tube is unstable and becomes turbulent owing to very small unavoidable disturbances. The first experiments of this kind were made by Reynolds on tubes of various diameters and with water of various temperatures. He found critical values of $\bar{u}r/\nu = 6,000$–$7,000$. His method, later also used by other investigators, consisted of letting a fine line of colored water flow into the test tube (Fig. 18). In the case of purely laminar flow, this thin colored line remains well defined, whereas in a turbulent flow it is disturbed and after some distance the water in the tube appears uniformly colored. Reynolds' expectation that the critical number could be made much larger by minimizing the disturbances has been found to be true by subsequent experiments of Barnes and Coker.[1] Their tests, conducted with

[1] BARNES, H. T., and E. G. COKER, The Flow of Water through Pipes, *Proc. Roy. Soc. (London)*, vol. 74, p. 341, 1905.

great precaution on a ¾-in. tube, led to a critical Reynolds' number of 10,000. An increase in temperature of the water of the storage tank causes convection currents and a consequent disturbance near the entrance of the test tube. This results in a considerable decrease in the critical Reynolds' number (at 75°C R_{crit} is about 5000).[1] Experiments by Ekman[2] conducted on Reynolds' original apparatus with the definite purpose of obtaining a high critical Reynolds' number led to values of $\bar{u}r/\nu = 20,000$ and in some isolated instances even up to 25,000. It seems, therefore, that the critical Reynolds' number does not tend

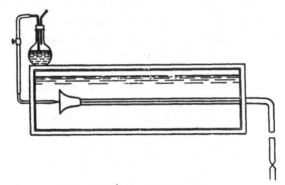

FIG. 18.—Reynolds' test set-up.

to a definite limit when the disturbances are made smaller and smaller, but that it can be made to exceed any value by increasing the precautions of the test.

At any rate, the experiments show that the critical Reynolds' number is a monotonous function of the initial disturbance, *i.e.*, the critical number always increases with a decrease in the disturbance. Whether there is an upper limit to the critical number with disturbances converging to zero is not yet known, but it does not seem to be very probable. On the other hand, there is a definite lower limit to the critical Reynolds' number at about $\bar{u}r/\nu = 1,000$ or somewhat above it. Below this, even very large initial disturbances are damped down, *i.e.*, below $R = 1,000$ a

[1] Other experiments by Barnes and Coker conducted on a 2¼-in. tube have not much meaning for the determination of the critical Reynolds' number, since the tube, being only 28 diameters long, was too short for the purpose (see Art. 25).

[2] EKMAN, V. W., On the Change from Steady to Turbulent Motion of Liquids (English), *Arkiv Mat. Astron. Fysik*, vol. 6, 1910.

turbulent flow with its typical irregular mixing motion and the consequent velocity distribution (Art. 34) cannot be maintained indefinitely.

With respect to the nature of the initial disturbances occurring in the tube, two questions have come up, namely: Which types of disturbance in the motion and which parts of the tube are of greatest importance for the creation of turbulence? The first of these questions has not been answered yet. As to the second, the entrance to the tube seems to be most sensitive to irregularities. Having taken care that the water in the tank has come to rest, which generally requires a few hours, it is found that in order to obtain a high critical Reynolds' number, it is especially important to round off the entrance of the tube. Very small irregularities in the shape of this first piece, where the boundary layer is as yet very thin, immediately cause a large drop in the critical Reynolds' number, while much larger irregularities at the wall of the tube far from its entrance hardly affect the critical Reynolds' number. For instance, Schiller[1] succeeded in obtaining the value R_{crit} = 10,000 with a tube of about ⅝-in. diameter in which a screw thread of about 1/64-in. depth had been cut, using however, a very well-polished, rounded-off entrance piece.

25. The Conditions at the Transition between Laminar and Turbulent Flow.—Now the phenomena in the range between the laminar and turbulent modes of flow will be discussed. Reynolds' original supposition was that when exceeding the critical velocity slightly, a weak turbulence would take place at first, which would become more violent with larger speeds. However, his experiments showed that no such gradual change occurs, but that the transition takes place very abruptly. Having a very definite colored line throughout the entire length of his tube just under the critical number, the slightest touch to the faucet would suddenly make it disappear. However in all cases laminar flow would persist in the first 20 or 30 diameters from the entrance of the tube, even when at greater distances the color would be completely mixed with the rest of the water. It has been supposed that the critical Reynolds' number might depend on the length of the tube. The fact that a complete turbulence takes place at such short distances from the entrance shows this supposition to be incorrect.

[1] SCHILLER, L., Roughness and Critical Reynolds' Number (German), *Z. Physik*, vol. 3, p. 412, 1920.

26. Intermittent Occurrence of Turbulence.—Reynolds made the observation that in many cases, especially with narrow tubes, the sudden destruction of the colored line did not occur over the total length of the tube but only in a part of it. Opening the faucet very carefully so as to reach the critical velocity gradually, the laminar flow would suddenly change into a turbulent one in a certain part of the tube, starting at about 30 diameters from the entrance, while still farther downstream the colored line remained visible. As soon as the turbulent mass of fluid, which was moving through the tube like a plug, had flowed out, a new turbulent region was formed at the same location.

The resistance to the flow for the total length of the tube increases when a part of the tube becomes turbulent; consequently, the mean velocity decreases, which brings it below the critical Reynolds' number. This phenomenon, which had already been observed by Hagen,[1] was studied in great detail by Couette.[2] He observed water flowing out of a large tank through the tube. The jet which came out of the end of the tube at first had a slightly rough surface and looked like a curved rod of frosted glass (turbulent flow). As the level of the tank went down, the jet became intermittently crystal clear and frosted, with a frequency which became faster and faster as the level came down. The clear jet would jump up, whereas the frosted jet would fall down so that a very regular oscillation took place. With the level of the tank sinking down still farther, the jet was clear most of the time and became frosted only once in a while. When Couette poured water into the tank, raising its level gradually, the same phenomena would take place in a reversed sequence. The surprising regularity of the oscillation of the jet in the region of the critical Reynolds' number can be judged from Fig. 19, where the mean velocity is plotted as a function of the time. This diagram was obtained by the author by means of moving pictures. In each individual picture the velocity was determined from the shape of the parabola of the jet. The maximum velocities correspond to those instants when turbulence suddenly starts in a part of the tube. Owing to the increased resistance in the tube caused thereby, the velocity decreases while turbulent flow exists. As soon as this turbulent "plug" starts

[1] See footnote on p. 29.
[2] COUETTE, M., Investigations on the Friction of Fluids (French), *Ann. chim. phys.*, vol. 21, p. 433, 1890.

to flow out of the tube, the resistance decreases again, which causes an increase in the velocity until the next maximum is reached and the phenomenon repeats itself.

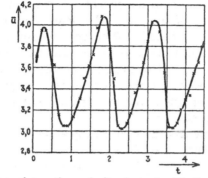

Fɪɢ. 19.—Variation of spouting velocity (m/sec) with the time (sec) in the critical region.

The phenomenon of intermittent turbulence can be explained somewhat differently by means of Fig. 20.[1] When starting with a condition below the critical Reynolds' number, determined by the point A, for instance, there will be a permanent laminar flow in the tube. When the velocity is slightly increased by opening the valve at the end of the tube until the critical value is exceeded (point B), turbulence will suddenly appear at a distance of about 30 diameters from the entrance. This turbulent plug of water is then pushed through the tube. The downstream end of this plug will move with the mean velocity $\bar{u}$, whereas the upstream end of it will move with a smaller velocity. This is due to the fact that at the upstream end new regions of

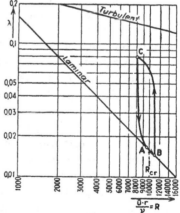

Fɪɢ. 20.—Variations of resistance coefficient in critical region.

turbulence grow continuously so that the length of the turbulent plug becomes gradually greater. The resistance coefficient

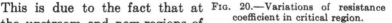

[1] The location of the starting of turbulence is rather close to the entrance, so that the points a and b do not lie on the straight line of Hagen-Poiseuille but slightly higher. This effect will not be considered here.

λ at any time depends on the ratio between the turbulent part and the laminar part of the tube. Its gradual increase causes a decrease in the mean velocity $\bar{u}$ and consequently in the Reynolds' number (B to C in the curve). When the turbulent plug of water flows out of the tube, λ becomes smaller again and the velocity increases (C to A in the curve), the point A in the curve corresponding to the instant when the turbulent part of the water has just left the tube. However, this condition cannot be steady since the pressure head in the reservoir is too high, which causes an acceleration of the water until point B is reached and the phenomenon repeats itself.

27. Measurements of Pressure Drop at the Transition between Laminar and Turbulent Flow.—In the investigation of the transition between laminar and turbulent flow by means of pressure-drop measurements, it is found that upon reaching the critical Reynolds' number the meniscus of the manometer, which had been completely quiet up to that time, begins to show irregularities. Having the faucet in a definite position, the meniscus moves irregularly up and down so that a reading of the pressure is hardly possible. If the precaution has been taken to make the measurement at a distance of at least 50 diameters away from the tank, it might be supposed that these irregularities are due to vortices in the tube which have not developed far enough to become a complete turbulence. This, however, is not so. Experiments have shown that the irregular condition of the meniscus is due to intermittent turbulence. On account of the large damping which usually exists in the manometer, it cannot follow the rapid oscillations between the laminar and turbulent states. Using a manometer with a very high damping (showing only mean values of the pressure over periods as large as ½ min), the meniscus varies gradually and not with jumps when passing from the laminar flow through the oscillating condition into the permanently turbulent flow. For this it is necessary to use a type of valve which allows very fine changes in the velocity instead of a common faucet.

28. Independence of the Critical Reynolds' Number of the Length of the Tube.—Schiller[1] has made the statement that the critical Reynolds' number depends on the length of the tube. Against his experiments, however, others can be brought up

---

[1] SCHILLER, L., Recent Experiments on the Problem of Turbulence (German), *Physik. Z.*, vol. 25, p. 541, 1924.

which prove the converse. Couette concluded from his experiments that the critical velocity above which intermittent turbulence takes place is independent of the length of the tube. Likewise, Barnes and Coker obtained the same critical Reynolds' number $R = 10,000$ with two tubes of the lengths 180 and 360 diameters respectively. Ekman, in his paper, defends himself against a possible objection that the very high critical Reynolds' numbers obtained by him might be due to the fact that the tubes used by him were so short that the small disturbances did not have a chance to grow into a complete turbulence. He reasons that if that objection were valid, the turbulence would have to start always at the end of the tube, coming nearer to the entrance with increasing velocity. This, however, is against experimental evidence. We mention again Reynolds' original experiments, who obtained the same critical number, 6,000, although the tubes being of the same length had diameters in the ratio $1:1.75:3.4$.

In this connection, a statement by Heisenberg[1] regarding the stability of fluid flow is of interest. He investigated the conditions under which a disturbance of the form $e^{i(\beta t - \alpha x)}\varphi(y)$[2] increases or decreases with the time, *i.e.*, whether the imaginary part of β, representing the negative damping, is positive or negative. From theoretical considerations he found that the negative damping is of the order of $(\alpha R)^{-\frac{1}{2}}$ and concluded from this that for very large Reynolds' numbers the negative damping becomes very small, so that the fluid under consideration has already left the tube when its instability would become serious. In this connection, it has to be considered that all quantities used in his paper have been made dimensionless, among others β. This, however, entails that the unit of time, with which the negative damping is measured, itself depends on the Reynolds' number, since the dimensionless t' is connected with the actual time t by the equation $t = \dfrac{\bar{u}}{r}t'$. When, therefore, the time t is measured in seconds and the length x in inches, it is found that the increase in the disturbance per second is of the order of

$$(\alpha r R)^{-\frac{1}{2}}\frac{\bar{u}}{r} = (\alpha r)^{-\frac{1}{2}}\left(\frac{\bar{u}\nu}{r^3}\right)^{\frac{1}{2}} = \frac{1}{\alpha^2}\frac{(\bar{u}\cdot\nu)^{\frac{1}{2}}}{r^2}$$

---

[1] HEISENBERG, W., On Stability and Turbulence in Fluid Flow (German), *Ann. phys.*, IV, vol. 74, p. 597, 1924.

[2] x is the coordinate in the direction of flow, y is perpendicular to it, and t is the time.

because

$$[\text{Imag. } \beta] = \frac{1}{\text{sec}} = \left[\frac{\bar{u}}{r}\right].$$

The increase in the disturbance per second therefore becomes larger for a tube of a given diameter when the velocity, and consequently the Reynolds' number, becomes larger.

C. TURBULENT FLOW

29. Historical Formulas for the Pressure Drop.—The pressure drop and the velocity distribution for the laminar flow through tubes of circular cross section can be derived from the differential equations. This, however, is not possible for turbulent flow. Figures 39 and 40, Plate 16, showing turbulent flow through long tubes make it plain that the motions are extremely complex. Therefore it seems hopeless to try to understand the mechanism of turbulence from the differential equations of Navier-Stokes. Figures 39 and 40 have been obtained by the author by photographing the surface of the water in a tank after having scattered aluminum powder on it. In Fig. 39 the velocity of the camera is very small so that the particles of aluminum near the walls of the channel appear as points. In Fig. 40, however, the velocity of the camera is about equal to the maximum velocity of the particles in the middle of the channel.

On the other hand, our interest in turbulent flow is much greater than that in laminar flow since the turbulent mode occurs much more frequently in nature and in technical applications. Therefore a great number of experiments have been carried out in order to determine the important relation between pressure drop and volume transported through tubes and channels. Comparing the results of these many experiments with the laws connecting pressure drop and mean velocity deduced from them, a very unsatisfactory picture is obtained. Nearly every investigator constructed his own pressure-drop formula from his experiments, largely owing to the fact that the similarity law of Reynolds was not known or at least was not used. Further no consideration was given to the roughness of the walls of the pipes or channels. This roughness, however, is of fundamental importance for the resistance of turbulent flow, as will be discussed later. For laminar flow it was seen before that the resistance is independent of the condition of the walls of the tube. The first formulas on resistance or pressure drop can be divided into three distinct classes, which have been discussed in detail by Hagen in 1869, namely:

$$\frac{\Delta p}{l} = \frac{C}{r} \cdot \bar{u}^2 \qquad [\text{Chézy (1775), Eytelwein (1822)}];$$

$$\frac{\Delta p}{l} = \frac{C_1}{r}\bar{u}^2 + \frac{C_2}{r^2}\bar{u} \qquad [\text{Prony (1804)}];$$

$$\frac{\Delta p}{l} = C\frac{\bar{u}^{1.75}}{r^{1.25}} \qquad [\text{Woltmann (1804), Flamant (1892)}].$$

In none of these formulas is the viscosity considered. This was done first by Reynolds, who plotted the pressure drop against the dimensionless number $\dfrac{\bar{u} \cdot r}{\nu}$. The resistance law deduced by Reynolds from his tests is

$$\frac{\Delta p}{l} = \text{const.}\left(\frac{r \cdot \bar{u}}{P}\right)^{1.723}\frac{P^2}{r^3} \qquad [\text{Reynolds (1883)}],$$

where P is a measure for the viscosity taken from Poiseuille's formula. The relation of P to the kinematic viscosity is expressed by the formula (c g s units):

$$\frac{\mu}{\rho} = \frac{0.01779}{1 + 0.03368T + 0.000221T^2} = 0.01779P,$$

where T is the temperature. The constant of Reynolds' formula, however, does not check with subsequent experiments; the pressure drop calculated by him comes out considerably too small.

According to this formula, the pressure drop is proportional to the 1.723 power of the velocity. However, the older measurements of Darcy which were plotted by Reynolds on a logarithmic scale showed that the exponent varies between the limits 1.79 to 2.00, depending on the material of the tube, *i.e.*, on the condition of the walls.

30. The Resistance Formula of Blasius for Smooth Tubes.— Based on Reynolds' law of similarity and a great number of tests up to $\bar{u}r/\nu = 50{,}000$ (especially those of Saph and Schoder[1]),

---

[1] SAPH and SCHODER, *Trans. Am. Soc. Civil Eng.*, vol. 47, p. 312, 1920.

Blasius[1] arrived at the following formula for the pressure drop
in smooth tubes:

$$\frac{\Delta p}{l} = \lambda \cdot \frac{1}{r} \cdot \frac{\rho \bar{u}^2}{2},$$ (6)

where

$$\lambda = \frac{0.133}{\sqrt[4]{R}}, \quad \left[R = \frac{\bar{u} \cdot r}{\nu} \right].$$ (7)

The main advantage of the pressure-drop law (1), as compared
with the older formulas, is the fact that for smooth tubes of the
same Reynolds' number (having different diameters, velocities,
and temperatures) the pressure drop expressed in units of
stagnation pressure per radius length is the same. Therefore
the complete resistance relation for smooth tubes can be expressed
by a single curve $\lambda = f(R)$. The fact that the points found from
experiments with different tube radii, velocities, and for different
fluids (water and air) lie on a smooth curve is to be interpreted
as an experimental verification of the law of similarity. The
agreement is so good that the scattering of the points is ± 2 per
cent at most.

Further measurements for higher Reynolds' numbers, especially those by Stanton and Pannell[2] and those by Jacob and
Erk[3] (up to about $R = 230,000$) show that the relation between
the resistance coefficient λ and R cannot be expressed by a simple
power for such a wide region of R. These experiments can be
better expressed by the formula

$$\lambda = 0.00357 + \frac{0.3052}{(2R)^{0.35}}$$

which was also found by Lees.[4]   According to measurements by

[1] BLASIUS, H. The Law of Similarity Applied to Friction Phenomena (German) *Physik. Z.*, vol. 12, p. 1175, 1911.   Or more in detail (German) *Forschungsarbeiten V. D. I.*, vol. 131.   The formula is also found in the book
by R. von Mises, "Elements of Technical Hydromechanics" (German),
Leipzig, 1914.

[2] STANTON, T. E., and J. R. PANNELL, Similarity of Motion in Relation
to Surface Friction of Fluids, *Phil. Trans. Roy. Soc. London*, (A), vol. 214,
p. 199, 1914.

[3] JACOB, M., and S. ERK, The Pressure Drop in Smooth Tubes and in
Standard Orifices (German), *Forschungsarbeiten V. D. I.*, vol. 267, 1924.

[4] LEES, *Proc. Roy. Soc.* (London) (A), vol. 91, p. 46, 1915.

Nikuradse[1] (up to about $R = 1.6 \cdot 10^6$) a systematic deviation from the curve of Lees is found for Reynolds' numbers greater than 200,000.

In the case of tubes of non-circular cross section or of open channels, it is not clear what is meant by the characteristic length in Reynolds' number. Considering that the most important property for the resistance of the tube is the ratio between the cross-sectional area A and the wetted periphery U, it is logical to take for the characteristic length the quantity

$$r = \frac{2A}{U}.$$

The reason for taking twice the area in the numerator is that in this way the quantity r becomes equal to the radius for a circular cross section. This quantity, which is in common use in hydraulics for open channels, is called the "hydraulic radius." In technical literature half this length $d = A/U$ can be often found referred to as the hydraulic radius. For very wide rivers, d is the mean depth. However in a non-circular cross section all parts of the periphery will not be equally important in creating resistance. Therefore it is necessary to get an experimental verification for the use of this hydraulic radius in the formulas. Experiments by Schiller,[2] Fromm,[3] and by Nikuradse[4] have shown that the influence of the shape of cross section is unimportant if the section is not too elongated. For laminar flow through rectangular and elliptical cross sections of various ratios of the axes, Boussinesq[5] has determined the influence of the shape of the cross section. For instance, he calculates $\lambda = 14.225/R$ for the square cross section, as compared with $\lambda = 16/R$ for the circular one.

31. The Resistance Law for Rough Tubes.—It was seen that the resistance relation up to relatively large values of R for smooth

[1] NIKURADSE, J., On Turbulent Flow of Water through Straight Tubes at Very Large Reynolds' Numbers (German), *Vorträge aus dem Gebiet der Aerodynamik*, edited by A. Gilles, L. Hopf, Th. von Kármán, Berlin, 1930.

[2] SCHILLER, L., On the Resistance to Flow in Tubes of Various Sections and Roughness (German), *Z. angew. Math. Mech.*, vol. 3, p. 2, 1923.

[3] FROMM, K., Flow Resistance in Rough Tubes (German), *Z. angew. Math. Mech.*, vol. 3, p. 339, 1923.

[4] NIKURADSE, J., Investigations on Turbulent Flow in Pipes of Various Cross Sections (German), *Ingenieur Archiv*. vol. 1, p. 306, 1930.

[5] BOUSSINESQ, J., Memoir on the Influence of Friction in Regular Fluid Motions (French), *J. math. pure et appl.*, vol. 13, p. 377, 1868.

tubes is completely given by the curves of Blasius and Lees. This is not the case for tubes with rough walls. The influence of wall roughness is always in the direction of increasing the resistance to turbulent flow; moreover, the various curves $\lambda = f(R)$ for different roughnesses do not coincide. This is due to the fact that Reynolds' law of similarity is not satisfied since for tubes of the same radius and different roughness or for tubes of different radius but the same roughness there is no geometrical similarity.

Blasius and von Mises[1] introduce a new quantity ϵ, proportional to the heights of the various roughness irregularities, and consequently make the resistance coefficient a function of ϵ/r, *i.e.*, the "relative roughness." Therefore

$$\lambda = f\left(R, \frac{\epsilon}{r}\right).$$

Blasius goes even farther than this. He does not consider the influence of roughness to be determined by the quantity ϵ, but he defines that two tubes of different radii have the same

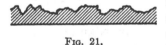

Fig. 21.

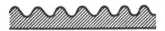

Fig. 22.

Figs. 21 and 22.—Examples of "roughness" of walls.

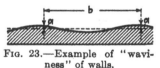

Fig. 23.—Example of "waviness" of walls.

roughness in case these tubes for some Reynolds' number give the same value of λ. If two tubes have the same roughness in this sense, the λ-values of both tubes can be represented by the same curve for all Reynolds' numbers. Therefore the resistance relation can be completely expressed by a family of curves depending on one parameter of which the curve of Blasius or Lees for smooth tubes forms a lower limit.

32. Roughness and Waviness of the Walls.—Measurements by Fromm and Schiller as discussed by Hopf[2] have shown that the resistance relations for rough tubes are more complicated. The law of resistance is affected not only by the relative magnitude of the various roughness irregularities but also by their form.

[1] See footnote on p. 42.

[2] HOPF, L., The Measurement of Hydraulic Roughness (German), *Z. angew. Math. Mech.*, vol. 3, p. 329, 1923.

According to Hopf and Fromm, there are two principally different types of roughness:

1. Roughness irregularities of short wave length and relatively high amplitude, as shown in Figs. 21 and 22. Examples of this are, for instance, cement walls, rusted steel, cast iron, corrugated steel. This type will be referred to as wall "roughness."

2. Very gradual irregularities of long wave length; for instance, planed wood, asphalted steel walls, as shown in Fig. 23. This type will be called wall "waviness."

For the first type of roughness the resistance coefficient λ is found to be nearly independent of the Reynolds' number but very much dependent on the "relative roughness" (Fig. 24).

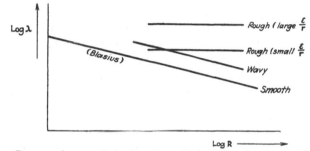

FIG. 24.—Pressure-drop coefficient *vs.* Reynolds' number for pipes with various kinds of walls.

This means that the resistance is proportional to the square of the velocity.

For the second type of roughness (waviness) larger values for the pressure-drop coefficient are obtained as compared with the smooth tube; however, the $[\lambda = f(R)]$-curve is parallel to the corresponding curve for the smooth tube while practically independent of the radius, especially for small ratios a/b of Fig. 23.

Measurements on drawn metal tubes with halfway smooth surfaces, being somewhere in the middle between roughness and waviness, show a gradual transition to the velocity-squared law with an increasing Reynolds' number. This can be interpreted as a confirmation of the remark by Schiller[1] that even for very smooth tubes the velocity-squared law will become true

[1] SCHILLER, L., The Problem of Turbulence and Connected Problems (German), *Physik. Z.*, vol. 26, p. 566, 1925.

for sufficiently large Reynolds' numbers. Since the phenomenon depends on the "relative roughness," it is to be expected that the velocity-squared law will be reached at smaller Reynolds' numbers for narrow tubes than for wide ones.

33. Measurement of the Mean Velocity of a Turbulent Flow by Means of a Pitot Tube.—The fact that the resistance coefficient for turbulent flow is materially greater than that for laminar flow is connected with the characteristic turbulent velocity distribution, which consists of a steep rise near the wall and a practically constant velocity over the rest of the cross section. In Art. 46 it will be seen that with certain assumptions it is possible to derive the turbulent velocity distribution from the law of turbulent pressure drop.

The velocity at a point of a turbulent flow is defined as the mean value of the velocity at this point with respect to time. If we denote by U, V, W the three rectangular components of the velocity in a point of a turbulent flow, it is possible to decompose these quantities into a part independent of the time u, v, w and a part giving the fluctuations with respect to time, u', v', w', *i.e.*:

$$U = u + u', \qquad V = v + v', \qquad W = w + w'.$$

The fact that such a procedure is possible, in other words, that experiments have shown that the mean values with respect to time of u', v', w' vanish for very short time intervals, shows that a turbulent flow is not quite without regularity. There are apparently laws determining this flow, although it seems that they can be approached only statistically.

The ordinary method of measuring velocity distributions by means of a Pitot tube and a fluid manometer gives mean values on account of the large damping usually existing in such instruments. Bazin has been the first to measure the turbulent velocity distribution across the circular section in this manner and has found for it the shape of a semi-ellipse with an axis ratio of 3.5:1. In the "immediate vicinity" of the wall, he found $u/\bar{u} = 0.741$, where $\bar{u}$ denotes the mean velocity with respect to the cross section. Stanton[1] made experiments on circular tubes, paying special attention to the steep velocity increase near the wall. To this end, the Pitot tube was sunk somewhat

[1] STANTON, T. E., The Mechanical Viscosity of Fluids, *Proc. Roy. Soc.* (London) (A), vol. 85, p. 366, 1911.

into the wall, which enabled him to determine the velocity at a distance of 0.001 in. from it.

It has to be considered that a damped manometer does not exactly indicate the mean value of u with respect to time but rather the mean value of the pressure difference which is proportional to U^2. Now in

$$U^2 = (u + u')^2 = u^2 + 2uu' + u'^2$$

the mean value of $2\,uu'$ is zero; however, the mean of u'^2 does not vanish. Since u'^2 is a positive quantity, the Pitot-tube indication of a pulsating velocity is always too high. For instance, if the variations in the velocity in short time intervals are of the order of ± 20 per cent, the reading of the manometer in terms of U^2 is equal to

$$u^2\left(1 + \frac{u'^2}{u^2}\right) = u^2\left(1 + \frac{20^2}{100^2}\right).$$

Extracting the root out of this expression, we get

$$u\sqrt{1 + \frac{u'^2}{u^2}} = u\sqrt{1 + \frac{400}{10,000}} = 1.02u,$$

or, in other words, our reading is 2 per cent high. Variations of ± 20 per cent in the velocity are rather large. In tubes or channels which are sharply divergent, such variations may occur;[1] with ordinary turbulence, however, the variations are considerably smaller. According to Burgers,[2] who measured the fluctuations of a turbulent stream of air by means of a hot-wire method, the fluctuations are less than ± 5 per cent. In that case, the error of the Pitot reading would be about 0.15 per cent.

34. The Turbulent Velocity Distribution.—Stanton also investigated whether the shape of the velocity-distribution curve depends on the Reynolds' number (the distribution being measured naturally at a sufficiently large distance away from the entrance of the tube). He found on smooth tubes that for the same Reynolds' number with differing diameters or velocities,

---

[1] KROENER, R., Experiments on Flow through Sharply Diverging Channels (German), *Forschungsarbeiten V. D. I.*, vol. 222.

[2] BURGERS, J. M., Experiments on the Fluctuations of the Velocity in a Current of Air (English), *Proc. Kon. Akad. Wetenschappen*, Amsterdam, vol. 29, No. 4.

the velocity distribution is always exactly the same. However, with varying Reynolds' numbers, the distribution curve changes, so that for larger Reynolds' numbers the velocity gradient at the wall becomes slightly steeper. This change in gradient becomes less and less with larger Reynolds' numbers, so that it may be concluded with reasonable assurance that the velocity distribution asymptotically reaches a certain limit for large Reynolds' numbers in smooth tubes. For tubes of considerable roughness (screw thread having been cut in), Stanton's measurements give independence of the velocity distribution from the Reynolds' number. This is tied up with the fact that for rough tubes the pressure-drop coefficient λ is a constant. Measurements of Fritsch[1] show that the influence of roughness of the walls on the velocity distribution is limited to the immediate neighborhood of the walls. He found that in comparing tubes of the same pressure drop with different roughness, the velocity distribution curve was the same in the center part of the section up to about 0.1 radius from the wall. In other words, the velocity-distribution curve depends only on the shear stress and not on the particular geometrical shape of the wall.

The change of the velocity-distribution curve with the Reynolds' number has some practical significance. In case such a change did not exist ($u_{max}/\bar{u}$ independent of R), it would be possible to determine the average velocity across the section by means of one single measurement of the velocity in the axis of the tube. This would mean a considerable simplification of the experimental procedure. Stanton and Pannell[2] have investigated the relation between $u_{max}/\bar{u}$ and R up to Reynolds' numbers of 42,000 with the result that $u_{max}/\bar{u}$ diminishes slowly with increasing R. Comparing the results of the various investigators, it is found that $u_{max}/\bar{u}$ asymptotically reaches the value 1.22 to 1.25 with increasing R.

35. The Turbulent Velocity Distribution in the Region of Transition near the Entrance of the Tube.—Just as in the case of laminar flow, there is a region of transition behind the entrance of the tube in which the final velocity distribution is formed. Experiments by Kirsten and Nikuradse show that

[1] FRITSCH, W., The Influence of Roughness on the Turbulent Velocity Distribution in Channels (German), *Z. angew. Math. Mech.*, vol. 8, p. 199, 1928.

[2] See footnote on p. 42.

the length of tube in which this occurs is materially shorter than in the case of laminar flow and, moreover, is less dependent on the Reynolds' number. Kirsten states that the transition takes place in a length of about 100 to 200 radii while Nikuradse's experiments show the final distribution at about 50 to 80 radii distance from the entrance. A still shorter transition length (probably too short) is given by the theory of Latzko.[1] According to him, with a Reynolds' number of 20,000 the final velocity distribution has been formed after about 20 radii.

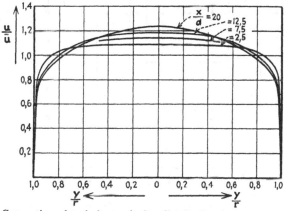

FIG. 25.—Generation of turbulent velocity distribution in entrance region after tests by Nikuradse.

Figure 25 shows the distribution at various distances from the rounded entrance as found experimentally by Nikuradse $\left(R = \dfrac{\bar{u} \cdot r}{\nu} = 25,000 \right)$. In Fig. 26 the ordinates of the curves of Fig. 25 have been plotted as a function of the distance from the entrance of the tube. For comparison the values as calculated by the theory of Latzko have been drawn in as dotted lines. It is seen that the curves even at a small distance from the entrance are radically different from those in the laminar case. In Fig. 27 the velocity distributions for the turbulent and laminar cases have been plotted at a distance of 5 radii from the entrance. It is seen that the turbulent curve shows an extraordinary rapid velocity rise near the wall of the tube. Therefore it is impos-

---

[1] LATZKO, H., Heat Transmission to a Turbulent Flow of Liquid or Gas (German), *Z. angew. Math. Mech.*, vol. 1, p. 268, 1921.

sible to talk about a laminar flow even at such a short distance
from the entrance as 5 radii.   However, the experiments of Reynolds with a colored line in the center of the tube show that

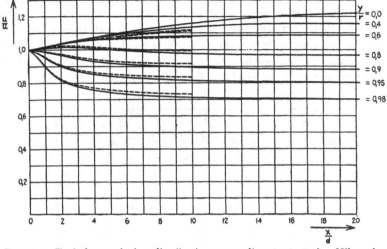

Fig. 26.—Turbulent velocity distributions according to tests by Nikuradse.

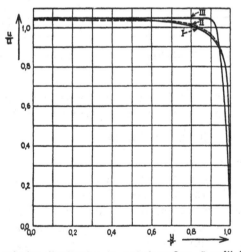

Fig. 27.—I, velocity distribution in turbulent flow, 5 radii from entrance;
II, the same curve calculated by the method of Latzko; III, velocity distribution
for laminar flow at same distance from entrance.

this line disappears only at a considerably greater distance from
the entrance.   It is possible to explain this by assuming that the
first turbulence does not appear in the middle of the tube but

rather at its walls. Figure 28 shows the final velocity distribution of turbulent flow in a circular tube at large distances from its entrance.

36. The Pressure Drop in the Turbulent Region of Transition. In the case of laminar flow, there is a considerable extra pressure drop in the entrance region. This is not the case in a turbulent flow, if the entrance has been rounded off sufficiently.

At the entrance of the tube, both kinds of flow experience a pressure drop of $\bar{u}^2/2g$ due to the conversion of the static head

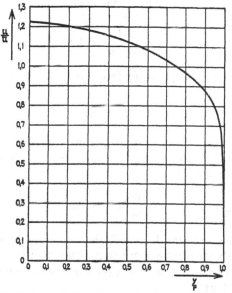

FIG. 28.—Turbulent velocity distribution in pipe far from entrance.

into velocity head. In the case of laminar flow, there is an additional pressure drop equal to that same amount since the kinetic energy of the parabolic velocity distribution is twice as large as the kinetic energy of the constant velocity distribution at the entrance. For the turbulent flow the additional pressure drop due to this effect is only $0.09\bar{u}^2/2g$.

If the pipe has a sharp-edged entrance, a vena contracta takes place. The contracted jet expands to the full pipe radius in a comparatively short distance causing an additional pressure drop. On page 243 of "Fundamentals,"[1] it was shown that the pres-

[1] See footnote, p. 3.

sure drop caused by a sudden widening of the cross section is equal to

$$h = \frac{1}{2g}(u_1 - \bar{u})^2,$$

where u_1 is the mean velocity at the smallest cross section A_1 and $\bar{u}$ is the mean velocity in the large cross section A. Introducing the contraction coefficient α,

$$\frac{u_1}{\bar{u}} = \frac{A}{A_1} = \frac{1}{\alpha}.$$

This becomes

$$h = \frac{\bar{u}^2}{2g}\left(\frac{1}{\alpha} - 1\right)^2.$$

Taking $\alpha = 0.64$, the pressure drop due to contraction becomes

$$h = 0.31\frac{\bar{u}^2}{2g}.$$

The total additional pressure drop therefore is made up of three parts, the entrance drop h_1, the acceleration drop h_2, and the jet contraction drop h_3:

$$h = h_1 + h_2 + h_3 = \frac{\bar{u}^2}{2g}(1 + 0.09 + 0.31) = 1.40\frac{\bar{u}^2}{2g}.$$

In Fig. 17 this amount $1.40\bar{u}^2/2g$ has been subtracted from the experimentally determined pressure drop, and the fact that such a smooth curve is obtained shows that this expression h is not far from the truth.

37. Convergent and Divergent Flow.—A very slight convergence or divergence of the walls of the tube or channel has a definite influence on the shape of the laminar flow. In the first place, the critical Reynolds' number determining the transition between the laminar and turbulent states is influenced considerably by a small deviation from parallelism of the walls. Secondly, the velocity distribution across the section and consequently the pressure drop vary considerably even with extremely small amounts of convergence or divergence.

A slight convergence has the tendency to stabilize the laminar mode of flow, *i.e.*, other conditions (shape of entrance, initial turbulence of water) being equal, the critical Reynolds' number increases considerably when the tube becomes slightly narrower in the direction of the flow. For slight divergence the conditions are opposite. Other things being equal, the turbulent mode of flow appears at considerably smaller Reynolds' numbers. However, exact numerical data for these phenomena are not available yet. Blasius[1] has calculated the change in the velocity distribution in the case of laminar flow through channels and tubes of varying width on the assumption that the divergence, *i.e.*, the angle between the wall and the axis, is small. Owing to the decrease in velocity, an increase in pressure appears which is superposed on the pressure drop due to friction. In case the resultant pressure drop in the direction of flow becomes negative, the possibility exists that the particles of the fluid in the neighborhood of the walls start moving backward. Denoting by $y(x)$ the shape of the divergent wall in the two-dimensional case and by R the Reynolds' number, the condition for beginning backward flow as found by Blasius is

$$R \cdot \frac{dy}{dx} = \frac{35}{4}.$$

A comparison, however, of this approximate result with the exact solution by Hamel,[2] leading to elliptic integrals, shows that only until about $R \cdot \frac{dy}{dx} = 3$ are the results of Blasius in satisfactory agreement with the exact solution. For diverging channels with straight walls (two dimensional) Pohlhausen[3] obtains the result that, even with vanishing divergence, a backward flow in the laminar boundary layer will occur as soon as the cross section has become about 22 per cent larger than the original section. However, the objection can be made that in the actual case laminar velocity distributions, with points of inflection as Pohlhausen finds them, can hardly occur. Among others, Rayleigh[4] has shown that motions with such points of inflection

---

[1] Blasius, H., Laminar Flow in Channels of Varying Width (German), *Z. Math. Physik*, vol. 58, p. 225, 1910.

[2] Hamel, G., Spiral Motions of Viscous Fluids (German), *Jahresber. deutsch. math. Ver.*, vol. 25, p. 34, 1916.

[3] Pohlhausen, K., Approximate Integrations of the Differential Equation of the Laminar Boundary Layer (German), *Z. angew. Math. Mech.*, vol. 1, p. 252, 1921.

[4] Rayleigh, Lord, On the Stability or Unstability of Certain Fluid Motions, *Proc. London Math. Soc.*, vol. 19, p. 67, 1887, or *Sci. Papers*, vol. 3, p. 17.

in the velocity distribution are very unstable. The great influence of a very small convergence or divergence on the velocity distribution of the laminar flow is shown in Fig. 29. At a Reynolds' number $R = 1,000$, the velocity distribution II occurs when the convergence of the tube is as little as $y = 1$ mm change in radius on a length of $x = 1$ m $\left(R\dfrac{dy}{dx} = 1 \right)$.

It is seen that the velocity distribution in the convergent tube is somewhat flatter than the parabola in the middle of the tube,

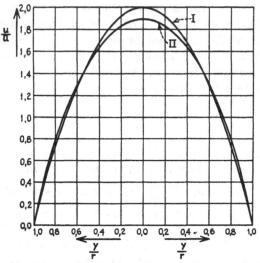

Fig. 29.—Laminar velocity distributions. I, constant cross section; II, Slightly converging pipe $R\dfrac{dy}{dx} = 1$.

while the velocity gradient at the wall has become somewhat greater. On the other hand, in the case of a divergent flow, there is a decrease in the velocity gradient at the wall while the velocity in the middle of the tube becomes somewhat steeper. The following consideration will make this plausible. In the case of a convergent flow, the mean velocity increases in the direction of the flow which causes an additional pressure drop besides the pressure drop due to friction. Considering two cross sections, 1 and 2, with the pressures p_1 and p_2 where $p_1 > p_2$, and denoting the velocity in some point of the first cross section by u_1 and the velocity of that point of the second cross section lying on the same streamline by u_2, Bernoulli's equation, neglecting friction, gives

$$p_1 - p_2 = \frac{\rho}{2}(u_2{}^2 - u_1{}^2)$$

or

$$u_2{}^2 = u_1{}^2 + \frac{p_1 - p_2}{\rho/2}.$$

In Fig. 30, u_1 and $\sqrt{\frac{(p_1 - p_2)}{\rho/2}}$ have been plotted as the sides of a right triangle. The hypotenuse of the triangle then represents the new velocity u_2, and, since $\sqrt{\frac{p_1 - p_2}{\rho/2}}$ is constant across

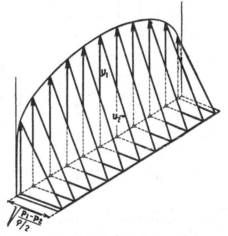

Fig. 30.—Graphical representation of the influence of convergence of the pipe on the velocity distribution diagram.

the entire section, it is seen that the velocity-distribution curve has become flatter. For divergent flows an analogous reasoning can be applied whereby it is only necessary to interchange u_1 and u_2 in Fig. 30.

Of the two possible modes of flow, the turbulent one is of greater practical importance, especially for divergent channels. For technical applications it is of great interest to know in which manner the energy loss due to a change from velocity head to pressure head depends on the angle of divergence of the tube, at which angle of divergence a back flow at the wall of the tube starts, in which location of a divergent flow the energy loss takes place, etc. However, these questions have not yet been answered satisfactorily. Beginnings of an answer can be found in the work by Gibson,[1] Andres,[2] Hochschild,[3] Kröner,[4] Dönch,[5] and Nikuradse.[6]

The experiments have been carried out partly with water and partly with air. In one of the investigations[3] an experimental proof of the mechanical similarity between air and water flow has been given.

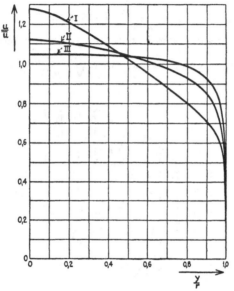

Fig. 31.—Turbulent velocity distribution in rectangular channel after F. Dönch. I, diverging channel (6 deg.); II, constant section channel; III, converging channel (5.8 deg.).

The cross sections of the tubes or channels used were mostly rectangular, the distance between the two small sides of the rectangle being kept constant. The influence of the conver-

[1] Gibson, A. H., *Proc. Roy. Soc. (London)* (A), vol. 83, 1910.

[2] Andres, Experiments on the Transformation of Velocity into Pressure with Water, *Forschungsarbeiten V. D. I.*, vol. 76.

[3] Hochschild, Experiments on the Flow in Divergent and Convergent Channels (German), *Forschungsarbeiten V. D. I.*, vol. 114.

[4] Kröner, R., Experiments on Flow in Sharply Diverging Channels (German), *Forschungsarbeiten V. D. I.*, vol. 222.

[5] Dönch, F., Turbulent Flow in Slightly Divergent and Convergent Channels (German), *Forschungsarbeiten V. D. I.*, vol. 282.

[6] Nikuradse, J., Experiments on the Flow of Water in Convergent and Divergent Channels (German), *Forschungsarbeiten V. D. I.*, vol. 289.

gence or divergence of the walls on the velocity distribution is principally the same as with the laminar flow; in this case also the distribution becomes flatter in the middle for converging walls and steeper in the middle for diverging ones. Figure 31 shows three such distributions as measured by Dönch.

When the angle of divergence becomes greater, backward flow at the walls takes place and the stream breaks away. This, however, does not take place symmetrically but always on one side only. The stream follows one of the walls but can be made to follow the other wall by very slight changes in the configuration. A disagreeable phenomenon in this connection (first observed by Kröner and later more extensively by Nikuradse) is that the two-dimensionality of the flow is destroyed. Even when the ratio of the sides of the entrance rectangle is as small as 1:8, the flow ceases to be two-dimensional before it breaks away from the wall at a diverging angle of 8 to 10 deg.

CHAPTER IV

BOUNDARY LAYERS

38. The Region in Which Viscosity Is Effective for Large Reynolds Numbers.—A consideration of the influence of inertia forces simultaneously with viscosity forces as in Oseen's theory (see page 264, "Fundamentals"[1]) is possible only for very viscous fluids or for very small Reynolds' numbers. In those cases, the "convection" terms of the acceleration become of importance only at very great distances from the body, where the velocity is hardly different from the velocity at infinity. Then Oseen's hypothesis can be used as a good approximation. In the immediate neighborhood of the body, however, where the velocity is very much different from the velocity at infinity, the flow is determined practically entirely by the action of viscosity, and the error made in the calculation of the inertia forces there is of little importance.

For large Reynolds' numbers (*i.e.*, large velocities or dimensions, small kinematic viscosity) the situation becomes entirely different. The inertia forces are then of much greater importance than the viscosity forces, at least at a sufficient distance from the walls or other obstacles, *i.e.*, with the exception of the layer of fluid near to the obstacle. However, if the influence of viscosity is completely neglected in the differential equation, erroneous results are obtained, as is discussed on page 104. This is due to the fact that the equations of Navier-Stokes in that case reduce to those of Euler where the boundary condition that the fluid does not slide along a solid wall cannot be satisfied (see page 260, "Fundamentals"[2]).

An important improvement in the treatment of fluid motions at great Reynolds' numbers, *i.e.*, for fluids of small viscosity, has been made by Prandtl.[3] His method will now be discussed.

[1] See footnote, p. 3.
[2] See footnote, p. 3.
[3] PRANDTL, L., On Fluid Motions with very Small Friction (German), *Proc. 3d Intern. Math. Cong., Heidelberg,* 1904.

Experience has shown that the motion of a fluid of small viscosity (water or air) around a body has velocities of the same order of magnitude as the velocity at infinity in practically the whole region with the exception of a thin layer surrounding the body. In the case of a streamlined body, as shown, for instance, in Fig. 32, the experimental streamlines and velocities coincide practically with those calculated on the basis of potential flow. More accurate experimental investigations of the velocity field have shown that right *at* the body, however, the fluid does not move relative to it. The transition from zero velocity to the velocity which can be observed near the body takes place in a very thin layer.

Therefore the field splits up into two regions:

1. Surrounding the surface of the solid body there is a thin layer where the velocity gradient $\partial w/\partial n$ generally becomes very large, so that even with very small values of the velocity w the

Fig. 32.—Cross section of streamlined body (strut).

shear stresses $\tau = \mu\dfrac{\partial w}{\partial n}$ assume values which cannot be neglected.

2. The region outside of this layer, where the velocity gradient does not become so large, so that the influence of viscosity is negligible. Here the streamline picture is entirely determined by the action of pressure, *i.e.*, it is the picture of a potential flow.

In general, it can be stated that the layer in which the velocity is reduced to zero owing to the action of viscosity is thin for small viscosities or, to be more general, is thinner the greater the Reynolds' number. Owing to this circumstance it is possible so to simplify the equations of Navier-Stokes for the thin boundary layer that an approximate solution becomes possible. The simplifications in this differential equation of the boundary layer are the more in agreement with experimental facts the thinner the layer is; or in other words, the solutions of the boundary-layer differential equation have an asymptotic character for Reynolds' numbers tending to infinity.

39. The Order of Magnitude of the Various Terms in the Equation of Navier-Stokes for Large Reynolds' Numbers.— Before proceeding to the simplifications in the equation of Navier-

Stokes (see page 258, "Fundamentals"[1]) for the boundary layer, we shall first discuss the order of magnitude of the various quantities appearing in the equation. For this purpose we shall consider conveniently a two-dimensional flow along a very thin flat plate as shown in Fig. 33. It will be found useful to employ dimensionless variables. The velocities will be expressed in terms of the velocity at infinity as a unit; the lengths in terms

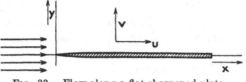

FIG. 33.—Flow along a flat sharpened plate.

of a characteristic length of the body, etc. The kinematic viscosity then is replaced by the reciprocal of Reynolds' number

$$\frac{\nu}{Vl} = \frac{1}{R}.$$

The x-component of the velocity u is supposed to be known outside the boundary layer and to be of the order 1. Assuming that the thickness δ of the boundary layer is small of the first order, we deduce from the identity

$$u \equiv \int_0^\delta \frac{\partial u}{\partial y} dy \sim 1$$

that the velocity gradient perpendicular to the plate $\partial u/\partial y$ is of the order $1/\delta$. This also can be seen by introducing the variable $\eta = y/\delta$ in the boundary layer, where η is of the same order as x. By this artifice, the coordinates are measured, as it were, with two different measuring sticks. We then have

$$\frac{\partial u}{\partial y} = \frac{1}{\delta} \frac{\partial u}{\partial \eta}$$

and

$$\frac{\partial^2 u}{\partial y^2} = \frac{1}{\delta^2} \frac{\partial^2 u}{\partial \eta^2}.$$

Since $\partial u/\partial \eta$ and $\partial^2 u/\partial \eta^2$ are of the order 1, it is seen that $\partial u/\partial y$ is of the order $1/\delta$ and $\partial^2 u/\partial y^2$ is of the order $1/\delta^2$.

[1] See footnote, p. 3.

Since, furthermore, $\partial u/\partial x$ is of the order 1, the continuity equation

$$\frac{\partial u}{\partial x} + \frac{\partial v}{\partial y} = 0$$

shows that $\partial v/\partial y$ is also of the order 1. With the aid of the identity

$$v \equiv \int_0^\delta \frac{\partial v}{\partial y} dy,$$

it is seen that v must be of the order δ; the same must be true of the quantities $\partial v/\partial x$ and $\partial^2 v/\partial x^2$, while $\partial^2 v/\partial y^2$ is of the order $\delta/\delta^2 = 1/\delta$.

Therefore the orders of magnitude of the various dimensionless terms in the two-dimensional equation of Navier-Stokes for the flow along a flat plate are as follows:

$$\frac{\partial u}{\partial t} + u\frac{\partial u}{\partial x} + v\frac{\partial u}{\partial y} = -\frac{1}{\rho}\frac{\partial p}{\partial x} + \frac{1}{R}\left(\frac{\partial^2 u}{\partial x^2} + \frac{\partial^2 u}{\partial y^2}\right) \qquad (1a)$$

$$1 \qquad\quad 1 \cdot 1 \qquad \delta' \cdot \frac{1}{\delta'} \qquad\qquad\qquad 1 \qquad\quad \frac{1}{\delta'^2}$$

and

$$\frac{\partial v}{\partial t} + u\frac{\partial v}{\partial x} + v\frac{\partial v}{\partial y} = -\frac{1}{\rho}\frac{\partial p}{\partial y} + \frac{1}{R}\left(\frac{\partial^2 v}{\partial x^2} + \frac{\partial^2 v}{\partial y^2}\right). \qquad (1b)$$

$$\delta' \qquad\quad 1 \cdot \delta' \qquad \delta' \cdot 1 \qquad\qquad\qquad\qquad \delta' \qquad\quad \frac{1}{\delta'}$$

In these equations δ' is the "dimensionless thickness of the boundary layer," *i.e.*, the thickness of boundary layer δ measured in terms of the characteristic length l or $\delta' = \delta/l$.

On the right side of Eq. (1a) $\partial^2 u/\partial x^2$ is small with respect to $\partial^2 u/\partial y^2$ so that it can be neglected. For the same reason $\partial^2 v/\partial x^2$ in Eq. (1b) can be neglected with respect to $\partial^2 v/\partial y^2$.

Inside the boundary layer the effects of the friction forces are of the same order as those of the inertia forces. The convective terms giving the effect of inertia on the left side of Eq. (1a) are of the order 1. Therefore it follows that $1/R$ is of the order δ'^2. Conversely, it follows that in a flow phenomenon where the viscosity is so small that in the fluid at large its action can be neglected with respect to that of inertia, a boundary layer is formed with a thickness $\delta' = \delta/l$ of the order

$$\sqrt{\frac{\nu}{ul}} = \sqrt{\frac{1}{R}}.$$

In order to get a quantitative picture of this, the following problem is presented: What is approximately the thickness of the boundary layer in the flow of Fig. 33 at a distance $l = 100$ cm from the sharp edge of the plate when the velocity at infinity is 100 cm/sec and the fluid is water of 20°C? (ν is 0.01 cm²/sec). The Reynolds' number then becomes $R = ul/\nu = 10^6$, and therefore δ' is of the order 10^{-3}. The thickness of the boundary layer $\delta = \delta'l$ is of the order $10^{-3} \times 10^2$ cm, i.e., of the order of 1 mm. In a layer of about this thickness the transition between the outside velocity and the velocity zero at the body is taking place.

40. The Differential Equation of the Boundary Layer.—Since in Eq. (1b) the various terms are of the order δ', $\partial p/\partial y$ must be of the same order. It is seen, therefore, that the influence of the y-dimension on the pressure inside the boundary layer can be neglected at least in the case of a thin boundary layer. In other words, in the layer the pressure is approximately equal to that in the outside flow so that in a sense the outside flow forces its pressure upon it. This result obtained from Eq. (1b) exhausts the information that can be had from that equation and hence it will not be considered any further.

Since inside the boundary layer p is a function of x only and independent of y, and since further $\partial^2 u/\partial x^2$ is negligible with respect to $\partial^2 u/\partial y^2$, the equation of Navier-Stokes for the boundary layer becomes

$$\frac{\partial u}{\partial t} + u\frac{\partial u}{\partial x} + v\frac{\partial u}{\partial y} = -\frac{1}{\rho}\frac{dp}{dx} + \frac{1}{R}\frac{\partial^2 u}{\partial y^2}. \tag{2}$$

Besides this equation, there is the continuity equation

$$\frac{\partial u}{\partial x} + \frac{\partial v}{\partial y} = 0,$$

which can be satisfied by introducing the stream function Ψ

$$u = \frac{\partial \Psi}{\partial y}, \quad v = -\frac{\partial \Psi}{\partial x}.$$

Then Eq. (2) becomes

$$\frac{\partial^2 \Psi}{\partial t\partial y} + \frac{\partial \Psi}{\partial y}\cdot\frac{\partial^2 \Psi}{\partial x\partial y} - \frac{\partial \Psi}{\partial x}\cdot\frac{\partial^2 \Psi}{\partial y^2} = -\frac{1}{\rho}\frac{dp}{dx} + \frac{1}{R}\frac{\partial^3 \Psi}{\partial y^3}. \tag{2a}$$

This differential equation of the boundary layer thus derived for the flow along a straight wall can also be written for curved boundaries; however, then it assumes a somewhat more complicated form.[1]

The boundary conditions to be imposed on Eq. (2a) are as follows: First, for $y = 0$, *i.e.*, at the boundary, we have

$$\Psi = 0, \frac{\partial \Psi}{\partial y} = 0.$$

Second, when y becomes of the order δ', the velocity in the boundary layer must asymptotically become equal to that of the outside flow or, since v in the boundary layer is neglected, u must become equal to $\bar{u}$, where $\bar{u}$ denotes the velocity parallel to the wall at a distance equal to the thickness of the boundary layer.

For instance, if the pressure distribution along the boundary of the body has been determined experimentally (see Art. 85), the velocity $\bar{u}$ can be calculated from Bernoulli's equation

$$\frac{\bar{u}^2}{2} = -\text{const.} \frac{p}{\rho}.$$

Therefore the total flow phenomenon of a fluid of small viscosity round the solid body is decomposed into a flow in a very thin layer where the internal friction has a definite influence and an outside flow where the viscosity has practically no effect. The pressure inside the boundary layer is determined by the flow outside it.

These statements are true only in case the boundary layer is actually sufficiently thin to warrant the simplifications made. This, however, is not always the case. Considering, for instance, the potential flow round a circular cylinder, we know that this solution gives a maximum pressure at the front stagnation point and a decreasing pressure from there along the sides of the cylinder extending to 90 deg. on either side of the stagnation point. The velocity along the wall rises to double the value of the velocity at infinity at the two points 90 deg. away from the stagnation point. From there on, a retardation of the fluid particles takes place until the rear stagnation point is reached. This is

---

[1] HIEMENZ, K., The Boundary Layer of a Straight Circular Cylinder in a Homogeneous Fluid (German), Dissertation, Göttingen, 1911; *Dinglers polytech. J.* (German), vol. 326, p. 321, 1911.

accompanied by a corresponding increase in the pressure.   Inside the boundary layer the individual fluid particles are always decelerated owing to the action of friction.   This retardation does not influence the outside flow very much as long as the particles are in the region of *de*creasing pressure.   However, in the part of the boundary layer where the pressure is *in*creasing, it may happen that the fluid particles, which have lost kinetic energy due to the action of friction, do not find it possible to overcome the pressure increase.   In the case of a *potential* flow, the kinetic energy accumulated is just sufficient to reach the rear stagnation point.   In the *actual* case, the particles in the boundary layer will come to rest in the region of increasing pressure and

Fig. 34.—Flow in boundary layer with increasing pressure in the direction of flow.

from then on they will experience an acceleration in the opposite direction owing to the pressure gradient.   The point of the boundary layer where this reversal in the motion takes place can be calculated only by an integration of the equation of the boundary layer as was done by Blasius[1] for the case of a circular cylinder.   The criterion determining the point where this back flow starts is

$$\frac{\partial u}{\partial y} = \frac{\partial^2 \Psi}{\partial y^2} = 0 \qquad (\text{for } y = 0),$$

since at the wall apparently $u = 0$ (see Fig. 34).

41. Definition of Thickness of the Boundary Layer.—The definition of the thickness of a boundary layer is arbitrary to a certain degree, since theoretically the transition of the velocity from zero to the potential velocity takes place asymptotically. In Fig. 35 the velocity distribution in the boundary layer is shown schematically for the case of two-dimensional flow past a plate, as discussed in Art. 39, the y-coordinate being exaggerated 1,000

---

[1] BLASIUS, H., Boundary Layers in Fluids of Small Viscosity (German), Dissertation, Göttingen, 1907, or *Z. Math. Physik*, vol. 56, p. 1, 1908.

times with respect to the x-coordinate. The thickness of the boundary layer may be defined arbitrarily as the distance from the plate where the velocity differs by 1 per cent from the velocity of the outer flow.

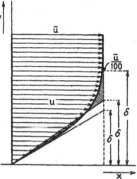

Another definition of the thickness of the boundary layer is obtained by taking the intersection of the asymptote and a straight line through the origin of the velocity-distribution diagram (Fig. 35) such that the shaded areas are equal. This thickness is somewhat smaller than the one defined first. A third possibility is the point of intersection of the asymptote with the tangent at the origin of the velocity-distribution diagram. This leads to a thickness which is only little smaller than that due to the previous definition.

Fig. 35.—Definition of thickness of boundary layer.

An entirely different manner of defining the thickness of the boundary layer is shown in Fig. 36 and expressed by the formula

$$\bar{u}\delta^{\star} = \int_0^{\delta} (\bar{u} - u)dy.$$

This thickness $\delta^{\star}$ therefore represents the amount by which the streamlines of the corresponding potential flow are shifted away from the boundary.

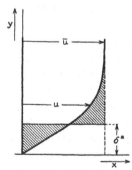

42. Estimate of the Order of Magnitude of the Thickness of the Boundary Layer for the Flow along a Flat Plate.— An estimate of the order of magnitude of the thickness of the boundary layer as was deduced in Art. 39 from the differential equation can be found for the flat plate also by means of a momentum analysis. The dotted line in Fig. 37 will be the boundary of the region to which we shall apply this analysis. It consists of a piece l starting at the front end of the plate, two straight pieces perpendicular to the points $x = 0$ and $x = l$, and finally a streamline which at the point $x = l$ has the distance δ

Fig. 36.—Definition of thickness δ^* of boundary layer.

from the plate. The momentum theorem, as discussed in Art. 100, "Fundamentals"[1] states that the flow of momentum through the bounding surface is equal to the sum of the pressure integral and the viscous force along the piece l of the plate. Since for the upper part of the bounding surface a streamline was chosen, no fluid passes through it. Consequently, the amount of momentum flowing through the two vertical parts of the boundary is the same. The mass of the fluid flowing through per second is approximately equal to $\frac{\delta}{2}bu$, where b is the width in the z-direction. This amount entering the left vertical part of the bound-

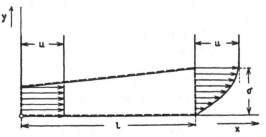

FIG. 37.—Application of momentum theorem for finding the order of magnitude of boundary layer thickness.

ary with a velocity u loses some of its velocity while flowing to the right so that a decrease in momentum takes place. The amount of this decrease is not known since in order to calculate it we have to know the velocity distribution at the point $x = l$ and consequently we have to know the thickness of the boundary layer, which is just what we want to find. However, it is possible to state that the change in momentum is proportional to $\rho\delta bu^2$. Since in a flow along a flat plate $dp/dx = 0$, the pressure integral taken on the entire boundary becomes zero. Therefore the decrease in momentum must be equal to the frictional force acting on the piece l of the plate. On page 4 it was shown that this force is proportional to $\mu lbu/\delta$. We have thus found that

$$\rho\delta bu^2 = C\mu lb\frac{u}{\delta}$$

or

$$\frac{\delta}{l} = C\sqrt{\frac{\mu}{\rho lu}} = C\sqrt{\frac{1}{R}},$$

[1] See footnote, p. 3.

where C is a numerical factor which cannot be found from this momentum consideration. The expression here derived for the flat plate is also valid for curved boundary layers in the steady state, which can be understood from the derivation given in Art. 39. For motions starting from rest (non-steady state), the relation valid for the first instant is

$$\delta = \sqrt{\nu t}.$$

The factor of proportionality C for the steady state can be calculated from the exact solution of Blasius.[1] Taking as a definition of the thickness of the boundary layer the intersection between the tangent at the origin and the asymptote, C was found to be 3.4, so that the thickness of the boundary layer along a flat plate becomes

$$\delta = 3.4\sqrt{\frac{x\nu}{u}} = \frac{3.4}{\sqrt{R}}x.$$

43. Skin Friction Due to a Laminar Boundary Layer.—By integrating the differential equation of the boundary layer, Blasius found for the shear stress $\tau_0 = \mu(\partial u/\partial y)_{y=0}$ the expression

$$\tau_0 = 0.332\sqrt{\frac{\mu\rho u^3}{x}}.$$

The frictional force along one side of a flat plate having the length l and the width b then becomes

$$D_f = b\int_0^l \tau dx = 0.332b\sqrt{\mu\rho u^3}\int_0^l \frac{dx}{\sqrt{x}} = 0.664b\sqrt{\mu\rho u^3 l}$$

or, introducing the drag coefficient c_f, defined by

$$D_f = c_f \cdot S \cdot \frac{\rho}{2}u^2,$$

where

$$S = bl = \text{surface},$$

this can be written

$$D_f = \frac{1.328}{\sqrt{R}} \cdot S \cdot \frac{\rho u^2}{2}$$

or

$$c_f = \frac{1.328}{\sqrt{R}},$$

a value which agrees very well with experiments on smooth surfaces.

[1] See footnote, p. 64.

Figure 38 shows the velocity distribution in the boundary layer along a flat plate as calculated by Blasius. The points indicated by little crosses in this figure were found experimentally by Hansen[1] by means of a very fine Pitot tube. It is seen that the agreement is very good.

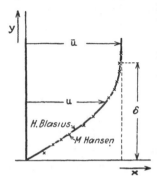

FIG. 38.—Velocity distribution in boundary layer along flat plate.

Based on some plausible assumptions regarding the velocity distribution in the boundary layer, von Kármán[2] gave an approximate procedure for calculating the thickness δ as a function of x and t for bodies of arbitrary shape by means of the momentum theorem. The procedure is valid only for very thin boundary layers. Having found the value of δ, the drag coefficient for any arbitrary body can also be calculated. In the analysis the pressure along the boundary of the body is supposed to be known. The calculations are started either by assuming the pressure distribution of the corresponding potential solution or by taking the pressures as found by experiment. The latter procedure gives a somewhat better agreement with the observed facts, as was shown by Hiemenz.[3] This method of von Kármán was carried through to numerical results for a few examples by Pohlhausen.[4] A very good agreement with the calculated values by Blasius was found. This is a distinct step forward since the results are found with very much simpler mathematical methods than the exact solution of Blasius, which involves complicated series developments.

44. Back Flow in the Boundary Layer as the Cause of Formation of Vortices.—The most important characteristic of the boundary layer is that under certain conditions a back flow takes place in it which leads to the creation of vortices and to a complete change in the flow pattern. This phenomenon will be illustrated later by photographs.

[1] HANSEN, M., The Velocity Distribution in the Boundary Layer of a Submerged Plate (German), Z. angew. Math. Mech., vol. 8, p. 185, 1928.

[2] VON KÁRMÁN, TH., On Laminar and Turbulent Friction (German), Z. angew. Math. Mech., vol. 1, p. 233, 1921.

[3] See p. 63.

[4] See p. 53.

An experiment on the two-dimensional flow through a diverging channel (Fig. 39) shows that in the first instant after starting a pure potential flow takes place, having a decreasing velocity in the direction of the flow owing to the increase in cross section. This decreasing velocity is accompanied by an increase in pressure, as follows from Bernoulli's equation, which means that some kinetic energy is transformed into pressure energy. A very short time after starting, however, the particles of the fluid in the thin boundary layer lose all their kinetic energy since they are slowed down, not only by the pressure gradient, but also by the friction forces. The particles thus coming to rest still experience the effect of the pressure gradient of the existing potential flow so that they now start to move backward. The flow in the boundary layer corresponding to this is shown in Fig. 39, in which for the sake of clearness the vertical y-ordinate has been very much exaggerated. The photographs of Figs. 24 to 33, Plates 12 to 14, show a corresponding phenomenon for the flow

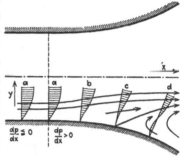

FIG. 39.—Boundary layers in a diffuser flow. The y-component perpendicular to the walls is greatly exaggerated.

past the tail end of a blunt body. In these pictures the flow is from left to right. Figure 24 shows the potential flow pattern, but in Fig. 25 it is seen that some fluid particles at the wall have come to rest. This fact is demonstrated by the particles appearing as sharp white points. The velocity diagram is approximately that of Fig. 39c. The next picture, Fig. 26 on plate 12 (being the third in a succession of exposures of a movie film), shows how these particles have taken a backward velocity from right to left and how at a certain distance from the body there is a line of fluid particles at rest relative to the body. Outside this line the original flow from left to right persists. This condition is illustrated approximately by the velocity distribution of Fig. 39d. In the following pictures of Plates 13 and 14, it is then seen how this dividing line of the forward and backward flows is unstable and breaks up into separate vortices. This finally causes a complete change in the flow pattern and consequently in the pressure distribution at the body.

45. Turbulent Boundary Layers.—It will be shown later that at a very high Reynolds' number the laminar boundary layer at the front part of the flow around a sphere becomes turbulent.

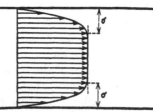

Fig. 40.—Laminar velocity distribution near entrance ($x/rR = 0.02$).

In the flow through a circular tube the jump from laminar to turbulent flow can also be considered as the transition of a laminar boundary layer into a turbulent one; in this case, the flow at the center of the tube follows the change also. On page 35 it was seen that the transition from laminar to turbulent flow in a tube always takes place with velocity distributions of the shape of Fig. 40. Here we have a laminar boundary layer at the wall of a cylinder which becomes turbulent when the critical Reynolds' number is reached. Experiments have shown that this turbulent boundary layer differs from the laminar one mainly in having a very much higher velocity gradient at the wall (Fig. 41).

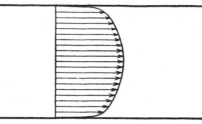

Fig. 41.—Turbulent velocity distribution.

46. The Seventh-root Law of the Turbulent Velocity Distribution.—Prandtl[1] has succeeded in finding an expression for the turbulent velocity distribution using only the resistance law for turbulent flow through smooth tubes as found by experiment. The assumptions on which his result is based are: (1) that the velocity distribution in the immediate neighborhood of the wall cannot depend on the radius of the tube but is determined completely by the quantities μ and ρ, as well as by the shear stress τ_0 at the wall; (2) that the velocity distribution curves remain similar with increasing velocity, i.e., when the maximum velocity

[1] PRANDTL, L., Investigations on Turbulent Flow (German), *Z. angew. Math. Mech.*, vol. 5, p. 136, 1925. See also von Kármán, footnote, p. 68.

in the center of the tube u_{max} is doubled, all other velocities are doubled also, so that

$$u = u_{max} f\left(\frac{y}{r}\right),\tag{3}$$

where y is the distance from the wall of the tube and r is its radius.

For a piece l of the tube, the relation between the pressure drop $p_1 - p_2$ and the shear stress τ_0 at the wall is

$$(p_1 - p_2)\pi r^2 = 2\pi r l \tau_0$$

so that

$$p_1 - p_2 = \frac{2l}{r} \cdot \tau_0.$$

On the other hand, Blasius' law for the pressure drop in smooth tubes (see page 42) is

(see page 42)

$$p_1 - p_2 = \frac{0.133}{\sqrt[4]{R}} \cdot \frac{l}{r} \cdot \frac{\rho}{2}\bar{u}^2.$$

Therefore the shear stress at the wall becomes

$$\tau_0 = \frac{0.033}{\sqrt[4]{R}}\rho\bar{u}^2 = 0.033\rho\nu^{\frac14}r^{-\frac14}\bar{u}^{\frac74}.\tag{4}$$

Now we specialize the general relation expressed by Eq. (3) in so far as we assume u to vary proportional to an unknown power q of the distance from the wall:

$$u = u_{max}\left(\frac{y}{r}\right)^q.$$

Writing $u_{max} = \text{const.} \times \bar{u}$ and eliminating $\bar{u}$ from Eq. (4), we get

$$\tau_0 = \text{const. } \rho\nu^{\frac14}r^{-\frac14}u^{\frac74}\left(\frac{r}{y}\right)^{\frac74 \cdot q}$$

or

$$\tau_0 = \text{const. } \rho\nu^{\frac14}u^{\frac74}\frac{r^{\frac74 \cdot q - \frac14}}{y^{\frac74 \cdot q}}.\tag{5}$$

The first assumption discussed above, stating that the shear stress at the wall is independent of the radius of the tube, requires that the exponent of r be zero, so that the following relation for q is obtained:

$$\frac{7}{4}q - \frac{1}{4} = 0$$

leading to

$$q = \frac{1}{7}$$

Therefore, based only on Blasius' experimental law of pressure drop and on the two assumptions discussed above, it is found that in a turbulent flow through smooth tubes the velocity increases as the seventh root of the distance from the wall:

$$u = u_{max}\left(\frac{y}{r}\right)^{\frac{1}{7}}.$$

In Fig. 42 the seventh power of the velocity as obtained experi-

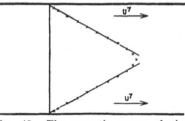

FIG. 42.—The seventh power of the velocity in a turbulent flow.

mentally is plotted against the distance from the wall. The fact that this curve comes out to be nearly a perfect straight line shows that the validity of the seventh-root law is not restricted to the immediate vicinity of the wall only but extends nearly to the center line of the tube. This could not have been anticipated.

In Art. 30 it was seen that Blasius' $\frac{1}{4}$th-power law breaks down for Reynolds' numbers above about 50,000, so that there the derivation of the seventh-root law has to be correspondingly changed. For instance, it has been found that for Reynolds' numbers of about 200,000, the velocity distribution near the wall is represented more nearly by an eighth-root law. For a Reynolds' number ten times as high again, we reach the tenth-root law. Prandtl[1] has shown that the above derivation can be generalized so as to find a velocity-distribution law for any experimental pressure-drop law that may come up.

Quite recently von Kármán[2] has shown that for theoretical reasons at very large Reynolds' numbers the expression $\dfrac{u_{max} - u}{\sqrt{\tau_0/\rho}}$ can depend only on y/r with the exception of a narrow region near the walls where the viscosity has a decided influence. The quantity $\sqrt{\tau_0/\rho}$ appearing in the denominator is a velocity of the same order of magnitude as the turbulent velocities u'

[1] See footnote, p. 70.

[2] VON KÁRMÁN, TH., Mechanical Similarity and Turbulence (German), *Nachr. Ges. Wiss.*, Göttingen, p. 58, 1930.

and v' discussed in Art. 33. Kármán's fundamental assumption is that the mechanism of turbulence at all locations in the fluid is of the same nature and differs only in the units of length and time employed. From this assumption he finds for the shear stress at any point outside the region near the wall,

$$\tau = k^2 \rho \frac{\left(\frac{du}{dy}\right)^4}{\left(\frac{d^2u}{dy^2}\right)^2}.$$

In this formula k is an empirical constant of a universal character. Since $\tau = \tau_0\left(1 - \frac{y}{r}\right)$, the formula for τ leads to a differential equation for du/dy which can be easily integrated twice and then gives an expression for $u(y)$. Von Kármán finds

$$\frac{u_{\max} - u}{\sqrt{\frac{\tau_0}{\rho}}} = \frac{1}{k}\left[\log \frac{1}{1 - \sqrt{1 - \frac{y}{r}}} - \sqrt{1 - \frac{y}{r}}\right].$$

For small values of y/r this becomes

$$\frac{1}{k}\left[\log \frac{2r}{y} - 1\right].$$

The agreement of this formula with the velocity distributions found at very great Reynolds' numbers is fairly good. The constant k turns out to be approximately 0.36.

The boundary between this region and the region near the wall in which the viscosity becomes important is determined by a definite value R_1 of the Reynolds' number $y_1\sqrt{\tau_0/\rho}/\nu$. The corresponding distance from the wall therefore becomes

$$y_1 = R_1 \frac{\nu}{\sqrt{\frac{\tau_0}{\rho}}}.$$

Denoting by u_1 the velocity corresponding to y_1, the above formula allows a calculation of $u_{\max} - u_1$. Since the mechanism of the flow in the region of viscosity is also the same all over, u_1 must be a multiple of $\sqrt{\tau_0/\rho}$;

$$u_1 = \beta\sqrt{\frac{\tau_0}{\rho}}.$$

Finally, therefore,

$$u_{\max} = \sqrt{\frac{\tau_0}{\rho}}\left\{\beta + \frac{1}{k}\left(\log \frac{2r\sqrt{\frac{\tau_0}{\rho}}}{R_1\nu} - 1\right)\right\}.$$

The relation between the shear stress τ_0 and the pressure-drop coefficient, λ, expressed in terms of the maximum velocity, is

$$\lambda = \frac{4\tau_0}{\rho u^2_{\max}}.$$

Using also a Reynolds' number in terms of u_{max},

$$R = \frac{r u_{max}}{\nu},$$

we get

$$\lambda = \frac{4k^2}{\{\log R\sqrt{\lambda} - \log R_1 + k(\beta - 1)\}^2},$$

or shorter

$$\lambda = \frac{4k^2}{\{\log R\sqrt{\lambda} + C\}^2}.$$

This formula gives a means for the calculation of λ for a given value of R by means of successive approximations, which converge very well. The mean velocity can also be calculated so that a comparison with the tests is possible. The agreement with the recent measurements of Nikuradse,[1] Schiller[2] and Hermann,[3] has been found to be very good in the region of large Reynolds' numbers. The value found for k was 0.44, and $C = 2.83$. Expressing the formula by means of $\bar{u}$ instead of u_{max} (with $R_m = \bar{u}r/\nu$), the equation still holds as an approximation formula:

$$\lambda_m = \frac{A}{\{^{10}\log (R_m\sqrt{\lambda_m}) + B\}^2}.$$

According to Nikuradse the values of the constants in this formula are $A = 0.133$ and $B = 0.18$.

47. Shear Stress at the Wall in the Case of a Turbulent Boundary Layer and the Thickness of This Layer.—The constant in Eq. (5) can be calculated numerically when it is taken into account that the maximum velocity in the center of the tube is between 1.22 and 1.25 times as great as the mean velocity $\bar{u}$. This is an experimental fact. Taking the average, i.e.,

$$u_{max} = 1.235\bar{u},$$

the expression for the shear stress at the wall, using Eq. (4), becomes

$$\tau_0 = \frac{0.033}{1.235^{1/4}} \cdot \rho\nu^{1/4}r^{-1/4}u_{max}^{7/4},$$

[1] NIKURADSE, J., Turbulent Flow of Water through Straight Tubes at very Large Reynolds' Numbers (German), *Vorträge Aachen*, 1929, p. 63, Berlin, 1930.

[2] SCHILLER, L., Pressure Drop in Tubes at High Reynolds' Numbers (German), *Vorträge Aachen*, p. 69, Berlin, 1929.

[3] HERMANN, R., "Experimental Investigation on the Resistance Law in Circular Tubes at Great Reynolds' Numbers" (German), Leipzig, 1930; HERMANN AND BURBACH, "Flow Resistance and Heat Transmission in Tubes" (German), Leipzig, 1930.

or introducing the seventh-root law, namely, $u = u_{max}(y/r)^{1/7}$, this is

$$\tau_0 = 0.0228\rho\nu^{1/4}u^{7/4}y^{-1/4}. \tag{6}$$

With the introduction of the dimensionless number uy/ν, the formula becomes

$$\tau_0 = 0.0228\rho u^2\left(\frac{\nu}{uy}\right)^{1/4}. \tag{6a}$$

As is evident from its derivation, this result is only true where the law of Blasius holds.

This formula, which is independent of the radius of the tube, can be applied also to other cases of turbulent flow along smooth walls, for instance, along a flat plate where there is a relatively thin turbulent boundary layer. In this case, the pressure along the plate can be considered constant as a first approximation (the same is true with the corresponding laminar flow). The influence of the viscous resistance is felt in an increase of the thickness of the boundary layer along the plate.

Prandtl[1] and von Kármán,[2] who invented this method of calculation independently of each other, assume a velocity distribution in the boundary layer expressed by

$$u = \bar{u}\left(\frac{y}{\delta}\right)^{1/7},$$

where $\bar{u}$ is the undisturbed velocity and δ is the thickness of the boundary layer. This assumption was made as an analogy to the flow through tubes. The shear stress at the wall can then be calculated:

$$\tau_0 = 0.0228\rho\bar{u}^2\left(\frac{\nu}{\bar{u}\delta}\right)^{1/4}. \tag{6b}$$

The total frictional resistance of one side of the plate of a length l and a width b then becomes

$$D_f = b\int_0^l \tau_0 dx.$$

This drag must be equal to the loss in momentum of the flow. In front of the plate each fluid particle has the velocity $\bar{u}$;

---

[1] PRANDTL, L., On the Frictional Resistance of Air (German), *Göttinger Ergebnisse*, vol. 3, p. 1, 1927.

[2] VON KÁRMÁN, see footnote, p. 68.

at the end of it this velocity has diminished to u. The corresponding mass flowing by per second is $\rho ubdy$, so that the loss in momentum becomes $b\rho\int u(\bar u - u)dy$. Substituting the above velocity distribution and performing the integration, this becomes

$$\frac{7}{12}\rho\bar u^2 b\delta.$$

Equating this expression to the drag where τ_0 assumes the value of Eq. $(6b)$, we get a relation from which δ can be immediately deduced. For convenience, we first calculate

$$\tau_0 = \frac{1}{b}\frac{dD_f}{dx}$$

which leads to

$$\frac{7}{72}\rho\bar u^2 \cdot \frac{d\delta}{dx} = 0.0228\rho\bar u^2\left(\frac{\nu}{\bar u d}\right)^{\frac14}$$

or

$$\delta^{\frac14}\frac{d\delta}{dx} = 0.235\left(\frac{\nu}{\bar u}\right)^{\frac14}.$$

By integration, we obtain

$$\frac{4}{5}\delta^{\frac54} = 0.235\left(\frac{\nu}{\bar u}\right)^{\frac14}x$$

or

$$\delta = 0.37\left(\frac{\nu}{\bar u}\right)^{\frac15}x^{\frac45} = \frac{0.37}{\sqrt[5]{\dfrac{\bar u \cdot x}{\nu}}}\cdot x. \tag{7}$$

Comparing this result with the corresponding expression for the laminar boundary layer on page 67, it is seen that the turbulent boundary layer increases with $x^{\frac45}$ while the laminar layer increases with $x^{\frac12}$. Therefore the turbulent boundary layer increases faster than the laminar one.

48. Friction Drag Due to a Turbulent Boundary Layer.— Continuing the calculations for turbulent flow, the relation between the shear stress and the length x along the plate becomes

$$\tau_0 = 0.0288\rho\bar u^2\left(\frac{\nu}{\bar u}\right)^{\frac15}\frac{1}{\sqrt[5]{x}}.$$

Consequently the drag for one side of a flat plate of length l and width b is

$$D_f = b \int_0^l \tau dx = 0.0288 \rho \bar{u}^2 \left(\frac{\nu}{\bar{u}}\right)^{\frac{1}{5}} b \int_0^l \frac{dx}{\sqrt[5]{x}}$$

or

$$D_f = 0.036 \rho \bar{u}^2 lb \left(\frac{\nu}{\bar{u}l}\right)^{\frac{1}{5}} = \frac{0.036}{\sqrt[5]{R}} \cdot \rho \bar{u}^2 lb,$$

where $R = \bar{u}l/\nu$. Introducing the familiar symbol λ for the drag coefficient, we obtain finally

$$D_f = \lambda S \cdot \frac{\rho}{2} \bar{u}^2 = \frac{0.072}{\sqrt[5]{R}} \cdot S \cdot \frac{\rho}{2} \bar{u}^2$$

with $S = bl$.
Figure 78 shows the curve

$$\lambda = \frac{0.072}{\sqrt[5]{R}}.$$

It has to be noted that the boundary layer at the front end of a well-sharpened plate is laminar at first and becomes turbulent at a certain critical Reynolds' number. According to the measurements of Van der Hegge-Zijnen, this happens at $R_{cr} = (\bar{u} \cdot \delta/\nu)_{cr} = 3,000$.

The deviations from Blasius' law mentioned in Arts. 30 and 46 are noticed in this case also for large Reynolds' numbers, from $ul/\nu = 5,000,000$ on. The derivations of Arts. 47 and 48 have been extended to this more complicated case by Schiller and Hermann.[1] Further, von Kármán[2] has extended his resistance theory of Art. 46 to the case of the flat plate. Both procedures lead to good agreement with the experiments.

The friction resistance of a rotating disk can be calculated in the same manner as the resistance of the flow along a plate at rest, as was shown by von Kármán in the paper cited on page 68. Let M be the moment or torque acting on the central circular part of an infinitely large rotating disk wetted on one side only. Further let r be the radius of this central part and U its circumferential speed; then we have for laminar boundary layers:

$$M = 1.84 r^3 \frac{\rho}{2} U^2 \frac{1}{\sqrt{R}}$$

[1] SCHILLER, L, and HERMANN, Resistance of Plates and Tubes with Large Reynolds' Numbers (German), *Ing. Arch.*, vol. 1, p. 391, 1930.

[2] VON KÁRMÁN, TH., Mechanical Similarity and Turbulence (German), *Proc. 3rd Intern. Cong.*, Stockholm, 1930, vol. 1, p. 85.

and for turbulent boundary layers

$$M = 0.146r^3\frac{\rho}{2}U^2\frac{1}{\sqrt[5]{R}},$$

where

$$R = \frac{Ur}{\nu}.$$

49. Laminar Boundary Layer inside a Turbulent One.—

Where in the previous discussions the terms "velocity" or "velocity distribution" were used in connection with turbulent flow, the mean value with respect to time was understood (see page 46). The actual velocity is found by superposing on this mean value the fluctuating flow which is characteristic of turbulence, amounting to approximately ±5 per cent of the mean velocity. The magnitude of these fluctuations must decrease rapidly when approaching the walls. This is particularly true for the component v perpendicular to the wall; the variations in the velocity u along the wall decrease much less rapidly; but in any case right at the wall the relation

$$\tau_0 = \mu\left(\frac{\partial u}{\partial y}\right)_{y=0}$$

is true for the mean values both of τ and of u. Assuming that the seventh-root law of the turbulent boundary layer is valid up to the wall itself, it follows that the shear stress becomes infinitely large since $\partial u/\partial y$ increases beyond all limits for $y = 0$. On account of this paradoxical conclusion the assumption is seen to be false. In fact the seventh-root law of the turbulent flow is valid up to a very small distance from the wall but ceases to be true exactly at it since there the transportation of momentum due to the turbulent fluctuations becomes zero. Therefore between the turbulent boundary layer with its seventh-root law and the wall there must be a very thin laminar boundary layer. Inside this latter layer the mean velocity gradient is found from the above equation for τ_0, where τ_0 is determined by Eq. (6). The part of the velocity-distribution diagram of Fig. 28 near to the wall has been plotted to a larger scale in Fig. 43 (up to $y/r = 0.1$, where y/r is measured from the wall, whereas it was measured from the center line in Fig. 28). It is seen that the seventh-root law approaches the wall with a tangent of zero angle. For large Reynolds' numbers the assumption is justified that the laminar boundary layer directly at the wall is very thin so that its velocity gradient can be approximately assumed to be a linear function of the distance

$$\tau_0 = \mu \frac{u}{y}.$$

With Eq. (4), this becomes

$$\tau_0 = \frac{0.033}{\sqrt[4]{R}} \cdot \rho \bar{u}^2, \tag{8}$$

$$0.033 R^{-\frac{1}{4}} \frac{\bar{u} \cdot r}{\nu} \cdot \bar{u} = u \frac{r}{y}$$

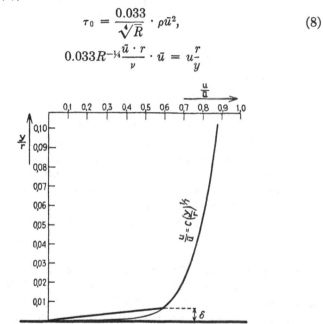

FIG. 43.—Laminar boundary layer at the wall for turbulent flow through pipe.

or introducing the seventh-root law and the relation $u_{max} = 1.235\bar{u}$

$$u = u_{max}\left(\frac{y}{r}\right)^{\frac{1}{7}} = 1.235\bar{u}\left(\frac{y}{r}\right)^{\frac{1}{7}},$$

$$\frac{y}{r} = \frac{68.4}{R^{\frac{7}{8}}}. \tag{9}$$

An approximate representation of the laminar boundary layer inside the turbulent one is found in Fig. 43, which corresponds to a Reynolds' number $R = \bar{u} \cdot r/\nu = 40,000$. At a distance from the wall $y/r = 68.4/40,000^{\frac{7}{8}} = 0.0065$, the seventh-root law is joined to the origin by a straight line. In the actual case, there is naturally no break in the velocity distribution curve, but a gradual transition. This is due to the fact that the turbulent fluctuations do not die out completely at a finite distance from the wall but rather diminish asymptotically toward it.

It is of interest to know the value of the velocity u in terms of the mean velocity $\bar{u}$ at a distance from the wall equal to the thickness of the laminar boundary layer.   Using the relations

$$\tau_0 = \mu \frac{u}{y} \text{ and } \tau_0 = \frac{0.033}{\sqrt[4]{R}} \cdot \rho \bar{u}^2,$$

we find

$$\frac{u}{\bar{u}} = 0.033 R^{3/4} \frac{y}{r},$$

or with Eq. (7)

$$\frac{u}{\bar{u}} = 0.033 R^{3/4} \cdot \frac{68.4}{R^{7/8}} = \frac{2.26}{R^{1/8}}. \tag{10}$$

In this connection, we refer to the early measurements of Stanton,[1] who from his experiments concluded the existence of a laminar boundary layer inside the turbulent one.

Although the thickness of these laminar boundary layers generally is extremely small, they can become of importance in the problem of heat convection due to flow past the body.   This will not be discussed here in detail, but the reader is referred to some papers by Prandtl[2] and to the experimental results of Schiller.[3]

50. Means of Avoiding the Creation of Free Vortex Sheets and Their Consequences.—It was seen in Art. 40 that back flow will take place in the boundary layer when there is a decrease in the velocity due to an increase in the pressure, as for instance in a diverging channel.   The result of this back flow, as shown in Figs. 24 to 31, Plates 12 and 13, is that a free vortex sheet gets into the fluid.   This sheet has been called "free" in order to distinguish it from the boundary layer, which is a vortex sheet clinging to the body.   The ultimate fate of such a free vortex sheet is to break up into individual vortices.

This formation of vorticity is undesirable, not only because it dissipates energy, but also because it changes the configuration of the flow so drastically that, in the divergent channel for

[1] See footnote, p. 46.

[2] PRANDTL, L., A Relation between Heat Convection and Flow Resistance in Fluids (German), *Physik. Z.*, vol. 11, p. 1072, 1910.   Also a Note on Heat Convection in a Tube (German), *Physik. Z.*, vol. 29, p. 487, 1928.

[3] SCHILLER, L., Investigations on the Problem of Heat Convection (German), *Z. angew. Math. Mech.*, vol. 8, p. 458, 1928.

instance, the desired pressure increase is practically prevented. The phenomenon can be avoided by making the angle of divergence of the channel (and consequently the pressure gradient) sufficiently small. In that case, the fluid outside of the boundary layer has sufficient power to pull the particles in the boundary layer with it against the small increase in pressure, so that back flow is avoided. This influence of the outer flow on the boundary layer is due to viscosity action in the case of laminar flow, while in the case of turbulent flow the transportation of momentum is the main agent pulling through the slow particles. The same principles apply to the flow around so-called streamlined bodies, such as, for instance, airships or airplane wings. Here, the pressure gradient at the tail is made sufficiently small by letting the body thin itself out very gradually. In this way a back flow in the boundary layer and the consequent formation of eddies can be avoided. The "pulling" action of the outer flow on the particles in the boundary layer again is much greater for turbulent boundary layers than for laminar ones on account of the interchanging of momentum of the individual particles of fluid inside the turbulent layer with those outside.

51. Influencing the Flow by Sucking away the Boundary Layer.—A method for preventing back flow and eddy formation at blunt bodies like spheres or cylinders, or in diverging tubes, was suggested by Prandtl[1] in 1904. It consists of sucking away into the interior of the cylinder those particles of the boundary layer that are just on the point of standing still before flowing back, thus preventing them from starting an eddy. The photographs of the flow round a circular cylinder with sucking on one side, published by Prandtl in 1904, show clearly how effectively the eddy formation is prevented. The power consumed by the process is relatively small, since very small volumes are involved, which was experimentally proved by J. Ackeret[2] for a diverging pipe and by O. Schrenk[3] for a sphere and several airfoils. Fig-

---

[1] See footnote, page 58.

[2] ACKERET, J., Sucking of the Boundary Layer (German), *Z. des V. D. I.*, vol. 35, p. 1153, 1926.

[3] SCHRENK, O., Experiments on a Sphere with Sucking Away of the Boundary Layer (German), *Z. Flugtech. Motorluftschiffahrt*, vol. 17, p. 366, 1926; Airfoils with Sucking of the Boundary Layer (German), *Luftfahrtforschung*, vol. 2, p. 49, 1928; Experiments on a Wing with Sucking of the Boundary Layer (German), *Z. Flugtech. Motorluftschiffahrt*, vol. 22, p. 259, 1931.

ures 36 to 38, Plate 15, show how drastically the flow pattern can be influenced by this method.   Figure 36 shows the flow from left to right through a very sharply diverging channel without any sucking.   It is seen, as was to be expected, that the fluid does not follow the walls of the tube at all, but breaks away from them at an early stage.   Figure 37, Plate 15, shows the flow which takes place when a small volume of the fluid is sucked away at the upper side, while the same phenomenon with sucking on both sides is shown in Fig. 38.   The flow in this picture is from left to right and is very nearly a potential flow where the kinetic energy is transformed into pressure energy according to the equation of Bernoulli.

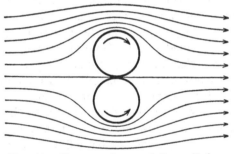

Fig. 44.—Flow round two rotating cylinders.

52. Rotating Cylinder and Magnus Effect.—Another method to prevent eddy formation, also initiated by Prandtl, consists of making the surface of the body move in the direction of the flow. This prevents a slowing down of the fluid particles at the surface and consequently prevents eddy formation.   The method offers greater practical difficulties than the one previously discussed in Art. 51.   For the simple case of two touching cylinders rotating in opposite directions the experimental difficulties are not great, the streamline picture being as shown in Fig. 44.

Not only was this case investigated by Prandtl (1906) but also that of a single rotating cylinder, which is of practical importance in connection with the "Flettner rotor."   The formation of eddies is avoided only on the side where the peripheral speed is in the same direction as the velocity of the outside flow, while on the side where these two velocities are opposite, the eddies develop so much easier.   The important feature of this flow is the fact that owing to the one-sided eddy formation the entire streamline picture becomes unsymmetrical. Figure 8, Plate 5, shows some moving pictures of the flow round a rotating cylinder starting from rest. The flow takes place from left to right and the rotation is clockwise; the ratio of the peripheral velocity u to the outside velocity v is kept constant equal to $u/v = 4$. Pictures 3 and 4 show particularly clearly how an accumulation of boundary-layer material is formed at the bottom side where the directions of the two velocities are in opposition. Owing to the small size of the pictures, the back flow inside the boundary layer is not visible. In picture 5 and the following ones, it is seen how this accumulation of boundary-layer fluid develops into a so-called "starting vortex." In accordance with the vortex

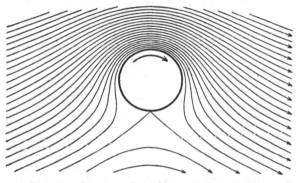

Fig. 45.—Flow round rotating cylinder; $u/v = 4$.

theorems of Helmholtz this vortex remains bound to the same fluid particles, so that in the end it is washed away with the fluid. The phenomenon is analogous to the starting of an airfoil, as described in Art. 99. Just as in that case the streamline picture, which remains after the starting vortex has been washed away, can be considered as the superposition of a potential flow round the cylinder and a circulation. The amount of circulation superposed has been taken such that in Fig. 45 the two stagnation points just coincide. A comparison of the constructed Fig. 45 with the photographic ones of Fig. 8, Plate 5, shows a very close agreement. Figure 46 gives the theoretical picture for the value $u/v = 2$; the corresponding photographs are shown in Fig. 13, Plate 8. For values u/v greater than 4, there will be a ring of fluid around the cylinder rotating about it permanently. Figure 9, Plate 6, shows the conditions for $u/v = 6$.

Just as in the case of the airfoil, there will be a lift on the rotating cylinder as soon as the starting vortex has been washed away. An inspection of the streamlines shows that they are crowded together considerably on top of the cylinder while on the bottom side they are spaced farther apart than normal. Consequently the velocity above the cylinder is larger than below, and an application of Bernoulli's equation (permissible outside the boundary layer) leads to an excess pressure below the cylinder and a lower pressure above it. In other words, there is a force perpendicular to the direction of the flow—a lift.

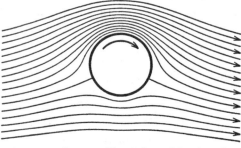

Fig. 46.—Same as Fig. 45 but with $u/v = 2$.

Since this lift is proportional to the circulation, it depends very much on the value of u/v. For $u/v = 4$ (Fig. 45) the two stagnation points just coincide, and the theoretical calculation gives for the lift

$$L = 4\pi\frac{\rho}{2}v^2ld,$$

so that

$$C_L = 4\pi \qquad \text{(see Art. 89)},$$

where d is the diameter and l the length of the cylinder. Measurements in the wind tunnel show a lift considerably smaller than this. The explanation for this discrepancy lies in the fact that the flow is not sufficiently two dimensional, as it was assumed in the calculation. Following a suggestion of Prandtl, this was corrected by providing the ends of the cylinder with disks of great diameter rotating with it. Figure 47 shows the so-called "polar diagram" (see Art. 90) of the cylinder with and without disks. It is seen that a rotating cylinder is capable of giving a much greater lift than an airfoil of the same projected area. However, this extra lift is dearly paid for with a drag several times greater than that of a good airfoil. Figure 48 shows the relation between the lift coefficient and the ratio of the peripheral and outside velocities for cylinders with and without disks.

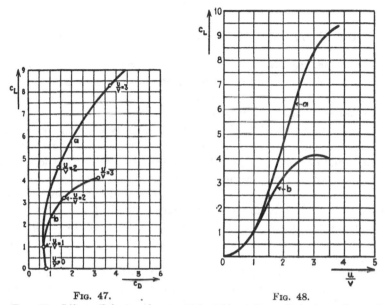

FIG. 47.   FIG. 48.

FIG. 47.—Lift coefficient *vs.* drag coefficient for rotating cylinder for various values of u/v; (a) with disks at the ends of twice cylinder diameter; (b) without disks.
FIG. 48.—Lift coefficient *vs.* u/v; same *a* and *b* as in Fig. 47.

In the next chapters on airfoils, yet another possibility of avoiding eddy formation will be discussed by which new energy is fed to the particles of the boundary layer which have been slowed down too much. In principle this is done by blowing air jets of great kinetic energy into the boundary layer through suitable nozzles. However the energy required for doing this is greater than that required for getting the same result by sucking some of the boundary-layer material into the inside of the wing.

CHAPTER V

DRAG OF BODIES MOVING THROUGH FLUIDS

53. Fundamental Notions.—When a body is moved with a uniform velocity along a straight line through a fluid at rest, it experiences a force in a direction opposite to that of the motion. This force is called the "drag" or "resistance."

Now the system of the fluid and of the body is given a uniform velocity opposite to that of the body. This brings the body to rest, while at infinity the fluid assumes a velocity equal and opposite to that velocity which the body had before. Since the superposition of such a uniform rectilinear motion on the system cannot have any dynamic consequences, the drag of a body is the same whether the fluid is at rest and the body is moving uniformly through it or whether the body is at rest and the fluid flows against it.

This presupposes that the individual particles of the fluid (at a sufficient distance in front of the body at rest) move completely uniformly and parallel to each other. This is not the case with natural fluid motions, as, for instance wind or water flowing through rivers. In Arts. 141 and 142 it will be seen to what extent this uniformity has been accomplished in artificial air motions in wind tunnels.

54. Newton's Resistance Law.—Historically the first resistance law was proposed by Newton, the founder of mechanics. With a small modification this law still holds today for motions where the drag is due to inertia, which often is the case for fluids of very small viscosity, like water or air. Newton's law is

$$D = fA\rho w^2,$$

where D is the drag, w the velocity, ρ the density of the fluid, and A the projected area of the body in the direction of the flow. The factor of proportionality f is as yet undetermined.

This law was derived by Newton from the momentum theorem: the force exerted by the fluid on the body is equal to the rate of change of momentum in the fluid due to the presence of the body.

Newton assumed instead of air or water a hypothetical medium of the following properties: the space round the body is filled with a large number of particles having mass but no length dimension. These particles which are at rest are not connected to each other, nor do they exert any influence whatever on each other. The body moving through this medium experiences impacts from all the particles in its path and consequently imparts momentum to them. The total mass of all the particles coming to impact with the body per second is $\rho A w$. This mass is given a velocity w' which is proportional to the velocity w of the body. The amount of momentum created per second, which has to be equal to the resistance or drag of the body, thus becomes

$$D = \rho A w w' = f A \rho w^2.$$

The resultant momentum depends on the assumption of whether the impact is elastic or non-elastic. Experiments indicated a more or less non-elastic impact. In the case of flow against a plane inclined under the angle α with respect to the direction of flow, Newton assumed this plane to be completely smooth. Thus only the component of the velocity perpendicular to the plane was annihilated. The mass per second affected is $\rho A w \sin \alpha$ and the loss in velocity $w \sin \alpha$ so that the force perpendicular to the plane becomes $\rho A w^2 \sin^2 \alpha$.

55. Modern Ideas on the Nature of Drag.—Newton's assumption led to a very simple formula for the proportionality factors f; however, it was found later that these factors did not coincide with the experimental ones. Newton's calculation for a square plane perpendicular to the direction of motion gives a factor 1, while the experiment leads to 0.55. The agreement is still worse for skew planes or for rounded bodies like spheres; while for streamlined bodies like those of airships, the drag is very much smaller than according to the Newtonian theory. The cause for this discrepancy is connected with the fact that Newton's assumption only takes into account the conditions at the front of the body, while those at the sides and at the tail end are left out of consideration. But just this tail end is of fundamental importance for the value of the drag.

The modern conception of the nature of drag is based more on the fact that the free paths of the individual particles or molecules are much too small than that the assumption of Newton could agree with experience. It has to be assumed therefore that the motion of a fluid particle is so influenced by its neighbors that neighboring particles stay together and occupy approximately the same amount of space during any short interval of time. The paths of the various particles therefore influence each other. This makes it clear that a calculation of the total drag by a simple summation of the actions on the various elements of the surface of the body, as in Newton's theory, is not admissible. The entire shape of the body is of importance for the force on any element of its surface on account of the mutual influence of the various fluid particles.

56. The Deformation Resistance for Very Small Reynolds' Numbers.—In the following discussion we assume that the body is moving in a straight line with a uniform velocity and is completely immersed in a homogeneous and incompressible fluid. Free surfaces do not occur and therefore the action of gravity is neutralized by static buoyancy. This leaves only viscous forces and inertia forces to act on the body.[1] In Art. 4 it was seen that the ratio of inertia forces and viscosity forces is given by the Reynolds' number, i.e., by an expression of the form wl/ν, where w is the velocity of the body, l some characteristic length of it, and ν the kinematic viscosity of the fluid. It is seen that for very large μ (sirup) or also for very small velocities or body dimensions (falling drops of a fog) the Reynolds' number can become very small. In such cases the influence of the viscosity forces on the geometry of the motion and consequently on the drag becomes of much greater importance than the influence of inertia. The body pushes itself through the fluid which is deformed by it. The resistance caused by this is due primarily to the forces necessary for the deformation of the various fluid particles. A system of stresses in the fluid is built up which transmits the force of the body to the fluid particles far away from it. In the case where there are solid walls in the neighborhood, these stresses are transferred to them. In an infinite extent of fluid, however, the force causes an acceleration of the total ocean of fluid. For very small Reynolds' numbers this deforming action takes place up to large distances from the body; for large Reynolds' numbers, however, it is restricted to the boundary layer. In the latter case the direct action of these viscous stresses is called the "skin friction" or "friction drag." In the "creep-

---

[1] The case where, due to free surfaces in the fluid, the gravity force is causing wave formation will be discussed in Art. 65.

ing" motion discussed here, this skin friction at the surface of the body and the pressure drag (both of which presuppose an inertia action) can be neglected with respect to the deformation resistance.

57. The Influence of a Very Small Viscosity on the Drag.— In most practical cases, however, the fluid is less viscous or the dimensions and velocities of the body are much larger. In such cases the inertia forces are very much larger than the viscosity forces. For instance, for a sphere of 2-in. radius moving with a velocity of 3 ft/sec in water, the ratio of the inertia forces to the viscosity forces is of the order of 50,000:1, since the Reynolds' number is approximately 50,000. The conclusion lies close at hand that under such conditions the influence of the viscosity on the drag can be completely neglected and that the drag is entirely determined by the inertia forces. However, a calculation of the drag due to inertia action in a completely frictionless fluid (which will be carried out in Art. 68) shows it to be zero. Since, therefore, inertia forces alone cannot be made responsible for the existing drag, it must be concluded that viscosity forces, however small they may be compared to the inertia forces, are necessary for the explanation of a drag. The great importance of a very small amount of viscosity in a fluid is due to the fact that it can completely change the picture of the flow in the ideal fluid (potential flow) on account of the formation of boundary layers and vortices (see Art. 44). Because of the influence of these, the potential pressure distribution at the surface of the body is changed to such an extent that the resultant of the pressure forces becomes different from zero. The component of this resultant in the direction of the motion is the drag. Indirectly, therefore, viscosity is the cause of the existence of drag.

In most practical cases, the inertia forces in the fluid at a certain distance from the body or the vessel walls will be very much greater than the viscosity forces, so that the latter can be neglected. However, as was explained in Chap. IV, this is not permitted in the thin boundary layer near the body. Since it is an experimental fact that fluid particles exactly on the surface of the body cannot flow past it, and since on the other hand the potential-flow theory of the ideal fluid leads to solutions with finite velocities along the surfaces, it follows that close to the surface of the body the viscosity forces become of the same order of magnitude as the inertia forces. The change in the velocity from the outer field to zero at the body itself takes place in a thin layer and consequently causes shear stresses in that layer. These stresses integrated over the entire area of the body make up the so-called "friction drag" or "skin friction."

For all practical cases, therefore, where the velocities are so great that the motion is not of the creeping type, the effect of viscosity on the drag of a body is twofold: (1) there are friction forces tangential to the surface of the body, the resultant of which is the friction drag; (2) the viscosity causes a change in the geometry of the streamline picture which in turn causes a change in the pressure field and consequently leads to a pressure drag.

58. The Relative Importance of Pressure Drag and Friction Drag with Various Shapes of the Body.—The shape and the position of the body determine to a great extent which part of the total drag is due to pressure and which is due to friction. In Fig. 49 a case is shown where the viscosity hardly makes a change in

Fig. 49.—Streamlined body.

the streamline picture, *i.e.*, a breaking away of the fluid from the body and its consequent eddy formation are prevented by the choice of a suitable shape. Because of this, the pressure drag is very small and of the same order as the friction drag. The pressure distribution in this case has a resultant which hardly differs from zero.

The converse is found, for instance, in the case of a flat plate moved in the direction perpendicular to its plane. Here the total resistance is almost entirely due to pressure while the friction drag can be neglected. The streamline pattern in front of the plate looks practically like the potential flow. On the back side, however, the viscosity forces have altered the shape of the velocity field completely, causing a corresponding change in the pressure.

On the other hand, if a flat plate is moved in the direction of its plane, there is hardly any influence of the viscosity forces on the potential-flow pattern. The resultant of the pressure forces is practically zero, so that the total drag, which is naturally smaller as in the previous case, is almost equal to the skin friction.

A subdivision of the total drag into its pressure and friction components can be effected experimentally by subtracting from the total resistance, as measured on the aerodynamic balance, the pressure drag as calculated from the pressure distribution, which in turn is obtained experimentally (see Art. 85). The difference thus found is the friction drag.

Formally the subdivision into friction drag and pressure drag can be accomplished by decomposing the total force on each element of the surface into a normal and tangential component. The normal components are to be interpreted as pressures, and their total resultant, or rather its component in the direction of the motion, is equal to the pressure drag. The tangential components are friction resistances, and their resultant in the direction of the motion is the friction drag.

For rough surfaces, it is convenient to carry out this decomposition into normal and tangential components for a smooth surface passing through the individual roughness irregularities. This leads to a subdivision into the two kinds of drags, but it is observed that the pressure drag on the individual roughness elevations in this case is calculated within the friction drag. The pressure drag depends to a great extent on the form of the body, while the friction drag is determined roughly by the area of its surface. This has led to the designations "form drag" and "surface drag." These names, however, are not very fortunate since the friction drag also depends somewhat on the form of the body.

59. The Variation of the Drag with Reynolds' Number.—It was seen that the total drag is composed of the three components: deformation drag,[1] friction drag at the surface of the body, and pressure drag, which is caused by the change in the geometry of the flow. For small Reynolds' numbers the drag consists primarily of the first kind, whereas for larger Reynolds' numbers it is made up of the two latter kinds. Which part of the total resistance is pressure drag and which is friction drag depends on the form and the position of the body. As generally stated, the

---

[1] The idea of "deformation resistance" for very small Reynolds' numbers as a part of what is usually denoted as friction drag is introduced here for the first time by the author. The work done by the deformation drag is ultimately dissipated into heat in the total field, including those parts at great distances from the body. The work of the friction drag in its more restricted sense and that of the pressure drag are dissipated into heat more specifically in the wake of the body.

drag depends on the shape and the position of the body, on its velocity, its size, and on the properties of the fluid. It is seen therefore that the complications of the problem of resistance in its most general form are so great as to leave little hope for a complete solution.

This is the reason that for the present most investigators have determined only the relation between the total drag and certain other physical quantities, without entering into details regarding the subdivision into the three kinds of drag. Since Newton's resistance law (discussed in Art. 54) in many cases agrees fairly well with the facts (although the underlying theory is wrong), it has become customary to write the drag formula in the form

$$D = \text{number} \cdot A\rho w^2,$$

where w is the velocity of the body relative to the undisturbed fluid, ρ the density of the fluid, and A the projected area of the body in the direction of the flow.[1]

Introducing the dynamic pressure $\rho w^2/2$, this drag can also be written

$$D = cA\frac{\rho w^2}{2}.$$

The factor of proportionality c, the "drag coefficient," is different for various shapes and various positions of the body. On the basis of the Newtonian conception of air resistance, it was thought for a long time that for a given shape and position of the body the drag coefficient was a constant (*i.e.*, independent of the size of the body and its velocity). It was thought therefore that for a given body the resistance law was completely known as soon as the drag coefficient had been determined for one single velocity. In particular, it was believed that with the drag coefficient thus determined the resistance of any other body geometrically similar to the test specimen could be calculated.

Experience has shown, however, that the conditions are much more complicated than this, which is not surprising after the

---

[1] Instead of taking the projected area in the direction of the flow, it is also possible to take any other characteristic area of the body or the square of a characteristic length of it. In the case of airfoils, it is customary to take the greatest possible projection. In case of comparison of drags of various airship bodies, one could also take $A = V^{\frac{2}{3}}$, *i.e.*, the square side of a cube having the same volume as the airship body.

discussions of the previous articles. The simple relations just outlined are practically true for bodies where the total drag consists almost exclusively of pressure drag and where the geometry of the flow, in particular the location of the breaking away of the fluid, is determined by sharp edges (as, for instance, with a plate perpendicular to the direction of the flow). In all other cases, however, where the resistance consists not only of pressure drag but also of friction drag or of deformation drag, the drag coefficient does depend on the size and velocity of the body. The reason for this is that geometrical similarity does not imply mechanical similarity and consequently does not imply similar flow patterns.

In Art. 4 it was seen that the condition for mechanical similarity, when geometrical similarity is insured, consists in having the same ratio between the inertia force and the friction force in similar points near the two bodies. In other words, mechanical similarity exists when the Reynolds' numbers for the two cases are the same. Only then are the drag coefficients for the two cases necessarily the same; a change in the Reynolds' number causes a change in the drag coefficient. This has been verified completely by experimental results. We have therefore

$$D = cA\frac{\rho w^2}{2} = f(R)A\frac{\rho w^2}{2}.$$

By using formally Newton's quadratic resistance law, all the complications of the various effects of the viscosity are expressed by the functional relation between the drag coefficient and the Reynolds' number. The knowledge of the law of similarity is a great advantage since, in carrying out the tests, it is necessary to vary only one parameter— for instance, the velocity. Then the dependence of the drag coefficient on the body dimensions and on the kinematic viscosity is automatically determined. A knowledge of the drag coefficient as a function of the Reynolds' number, therefore, enables us to calculate the drag of a certain shape for all fluids, all velocities, and all body dimensions. The relation itself, however, can be ascertained only by experiment, and for each shape and position of the body the experiment has to be repeated. In other words, to each shape and position of the body there belongs a characteristic function $c = f(R)$.

60. The Laws of Pressure Drag, Friction Drag, and Deformation Drag.—First some general statements on the subject will be made:

1. In case the total drag of a body consists almost exclusively of pressure drag, the function $f(R)$ is practically a constant (except for small Reynolds' numbers):

$$D = \text{const. } A\frac{\rho w^2}{2}.$$

This is the case, for instance, with plates moved perpendicularly to their plane or generally for bodies with sharp edges where the fluid breaks away at definitely determined points of the body.

2. On the other hand, if the total resistance is practically entirely due to friction, as with a plate moved along its plane, two different resistance laws have to be distinguished, according to whether the Reynolds' number $R = lw/\nu$ (l is the length of the plate in the direction of the flow) is smaller or larger than about 5.10^5. It is assumed that the surface of the plate is smooth.

a. For values of R smaller than about 5.10^5, Blasius[1] derived a formula on the basis of Prandtl's boundary-layer theory. He found that c is inversely proportional to the square root of the Reynolds' number, so that the resistance law appears in the form

$$D = \frac{1.327}{\sqrt{R}} \cdot S\frac{\rho w^2}{2} \quad \left(\text{for } R = \frac{wl}{\nu} < 5 \cdot 10^5\right)$$

where S is the total surface of the plate.

b. For Reynolds' numbers greater than 5.10^6 (*i.e.*, from about ten times the limit mentioned above), experiments of Wieselsberger[2] and Gebers[3] indicate that the drag coefficient is proportional to the reciprocal of the fifth root of Reynolds' number.

$$D = \frac{0.074}{\sqrt[5]{R}} \cdot S\frac{\rho w^2}{2} \quad \left(\text{for } R = \frac{wl}{\nu} > 5 \cdot 10^6\right)$$

This change in the resistance law is caused by the fact that for the lower range of Reynolds' numbers the flow along the plate is laminar, while it becomes turbulent for the higher range (see Art. 63).

[1] BLASIUS, H., Boundary Layers in Fluids of Small Viscosity (German), Z. Math. Physik, vol. 56, p. 1, 1908.

[2] WIESELSBERGER, C., Investigations on the Frictional Resistance of Canvas Covered Plates (German), Göttinger Ergebnisse, vol. 1, p. 121, 1925.

[3] GEBERS, Note on the Experimental Determination of the Resistance of Bodies Moved in Water (German), Schiffbau, vol. 9, 1908.

c. In the intermediate region a transition between the two resistance laws takes place, which, according to L. Prandtl,[1] can be represented by the formula (see Art. 86)

$$c = \frac{0.074}{\sqrt[5]{R}} - \frac{1,700}{R} \qquad \text{(for } 5 \cdot 10^5 < R < 5 \cdot 10^6\text{)}$$

3. For very small velocities or dimensions or for very viscous fluids (Reynolds' number smaller than one) Stokes's law (Art. 261, "Fundamentals"[2]) gives proportionality of the drag with the first power of the velocity and the first power of the length dimension. In terms of Newton's resistance law, this can be formally expressed as a proportionality with the reciprocal Reynolds' number:

$$D = \text{const. } \mu l w = \frac{c}{R} A \frac{\rho w^2}{2} \qquad \text{(for } R \ll 1\text{)}$$

where l is a characteristic length dimension of the body. It is found, therefore, that the deformation resistance is not proportional to the square but to the first power of the velocity.

61. General Remarks on the Experimental Results.—During the last half century a great number of resistance measurements in air and water have been carried out, and an extensive literature on the subject has been accumulated.[3] However, the measurements of the last 20 years have been of more importance than the earlier ones. The cause for this is that the older experiments were practically all based on Newton's impact theory of the resistance, according to which the drag is determined solely by the shape of the front side of the body. Only after it was appreciated that the flow around a body has to be interpreted as a flow of a continuum, the necessary conditions for properly conducting drag measurements were recognized. In particular, it had to be appreciated that there should be left sufficient space round the body so that the fluid can flow past it in an undisturbed manner. Also it is necessary to take care that the flow is not disturbed by the presence of other obstacles situated either to the side or even behind the body under test. The older measurements, which neglected these conditions more or less, show considerable spreading of the experimental data, whereas

[1] See footnote, p. 75.
[2] See footnote, p. 3.
[3] A good bibliography up to the year 1910 can be found in the book by G. Eiffel, "The Resistance of Air" (French), Paris, 1910.

the modern experimental results agree very well among themselves. With modern experiments it is also found that the drag of geometrically similar bodies at the same Reynolds' number is the same whether the body is moved through a fluid at rest or whether the body is at rest and the fluid is flowing against it (if sufficient care is taken that the fluid is moving uniformly, without too much vorticity). In the following articles the experimental drag coefficients will be given as functions of the Reynolds' number for various shapes of the body.

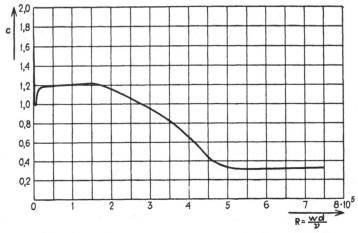

Fig. 50.—Drag coefficient *vs.* Reynolds' number for two-dimensional flow round circular cylinder.

62. The Relation c = f(R) for the Infinite Cylinder.— Figure 50 shows this relation for a cylinder of infinite length with its axis perpendicular to the direction of the flow (*i.e.*, the two-dimensional case). The region of Reynolds' numbers of practical interest is enormously wide, in this case up to about $8 \cdot 10^5$. In the ordinary manner of plotting, the region of the smaller Reynolds' numbers (up to about $R = 10,000$) shrinks together so much that no detail of the curve can be recognized. In order to circumvent this difficulty, it has been found practical to plot the logarithms of the drag coefficients as well as of the Reynolds' numbers instead of these quantities themselves. Another way of doing the same thing is to plot the drag coefficient and the Reynolds' number itself on logarithmic paper. Figure 51 shows the curve of Fig. 50 replotted in this manner. It was determined by C. Wieselsberger by measuring the resistances of a number of cylinders of various diameters (from 0.002 in. to 12 in.) at various velocities. From these experiments the drag coefficient was determined by the usual formula

$$c = \frac{D}{A\frac{\rho w^2}{2}}.$$

The fact that the drag coefficients of cylinders of various diameters lie on a single curve is to be interpreted as an experimental verification of the law of similarity.

In the region of R between 16,000 and about 180,000, the quadratic resistance law is found to hold with good accuracy, the

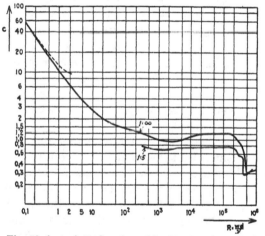

Fig. 51.—Like Fig. 50, but plotted on logarithmic paper for cylinder of infinite length and for one of five-diameters length. (*Wieselsberger*.)

drag coefficient being practically constant = 1.2. With decreasing Reynolds' numbers the drag coefficient first decreases,[1] but with a still further decrease of R the value of c increases again. The smallest value of R for which the resistance coefficient was determined is 2.1. For very small values of R ($R \ll 1$), Lamb[2] has derived a formula for the drag of a cylinder, assuming preponderance of the viscosity action. This relation between

[1] Relf, E. F., Discussion of the Results of Measurements of the Resistance of Wires, *Repts. and Mem. Nat. Adv. Comm. Aeronautics* (*London*), p. 47, 1913–1914.

[2] Lamb, H., On the Uniform Motion of a Sphere Through a Viscous Fluid, *Phil. Mag.*, vol. 21, p. 120, 1911.

the drag coefficient and R is drawn in the curve as a dotted line. It is seen that an extrapolation of the experimental curve fits the theoretical one of Lamb without any trouble.

63. The Region above the Critical Reynolds' Number.—We shall now discuss a very peculiar phenomenon in the resistance law: in the region of Reynolds' numbers between $2 \cdot 10^5$ and $5 \cdot 10^5$,

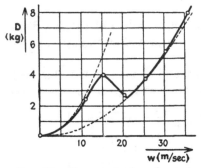

FIG. 52.—Decrease in drag due to increasing velocity of cylinder. (*Wieselsberger.*)

the curves of Fig. 50 or 51 suddenly drop from $c = 1.2$ to $c = 0.3$. This decrease in the drag coefficient is so large that the drag itself, instead of increasing quadratically with the velocity, even *decreases* with increasing velocity. Figure 52, for instance, shows the resistance for a cylinder of 12-in. diameter and a length of 40 in. It is seen that the resistance drops from 4 kg at 15m/sec to about 2.5 kg, in spite of the fact that the velocity increases to about 21m/sec.

This fact was first discovered by Constanzi[1] for spheres in water and by Eiffel[2] for spheres in air. Prandtl[3] observed it subsequently and found an explanation in the fact that, in surpassing a certain velocity, the boundary layer at the front end of the sphere experiences a definite change. Below this so-called "critical Reynolds' number," the flow in the boundary layer is laminar, while above this Reynolds' number it suddenly becomes turbulent.

A more detailed investigation shows that the mechanism is as follows: The eddies in the turbulent boundary layer cause small quantities of the fluid in the dead-water region (near the point of breaking away of the flow) to be pushed backward. This causes the point of breaking away to be moved backward along the body, which makes the total eddying dead-water region

[1] CONSTANZI, G., Some Experiments in Hydrodynamics (Italian), *Rendiconti della esperienze e studi nello stab. di esp. e costr. aeronautiche del genio,* vol. 2, p. 169, Rome, 1912.

[2] EIFFEL G., On the Resistance of Spheres in Air (French), *Compt. rend.,* vol. 155, p. 1597, 1912.

[3] PRANDTL, L., Air Resistance of Spheres (German), *Nachr. Ges. Wiss. Göttingen, Math.-physik. Klasse,* 1914.

become smaller. Since the pressure drag of a body is determined primarily by the kinetic energy in the eddies of the wake, it is clear that a decrease of the size in this region causes a smaller resistance.

Detailed investigations[1] have shown that for round bodies without sharp edges the manner of support of the test body in the air current and the state of turbulence of the air are of great importance. In case the body is supported by means of a strut attached to it in the dead-water region, and not by means of thin wires in the streamline region, the boundary layer is not disturbed and it is possible to obtain a greatly diminished drag in the region above the critical Reynolds' number without influencing the actual value of the critical Reynolds' number itself. On the other hand, if the air is made very turbulent (for instance, by putting a net of thin wires in front of the sphere), or if the flow in the boundary layer is affected greatly by putting a very thin wire around the body,[2] the critical Reynolds' number, at which the drag coefficient suddenly drops, is found to be much smaller. Prandtl has proposed to use the critical Reynolds' number for a smooth sphere with a definite support as a measure for the uniformity of an air stream.

64. The Resistance Law for Finite Cylinders, Spheres and Streamlined Bodies.—Figure 51 shows the resistance curves for an infinite cylinder and for a cylinder of 5 diameters length. In the latter case the flow is three dimensional, causing a considerable decrease in the drag coefficient as compared to the case of two-dimensional flow. Figure 53 shows the drag coefficient for a Reynolds' number of $8.8 \cdot 10^4$ as a function of the ratio between the length l and the diameter d. It is seen that the resistance coefficient of a cylinder of 1 diameter length ($d/l = 1$) is about half as large as for one of infinite length ($d/l = 0$). This phenomenon, which can be observed also with other bodies (for instance, with plates of various ratios between the sides), is due to the fact that in the three-dimensional flow the fluid can leak around the flat ends of the cylinder into the dead-water region, *i.e.*, into the region of low pressure. This causes a different pressure distribution, leading to a smaller pressure drag. The

[1] FLACHSBART, O., New Experiments on the Air Resistance of Spheres (German), *Physik. Z.*, vol. 28, p. 461, 1927.

[2] WIESELSBERGER, C., The Air Resistance of Spheres (German), *Z. Flugtech. Motorluftschiffahrt*, vol. 5, p. 140, 1914.

relative importance of this side leakage for long cylinders as compared with shorter ones can be seen in Fig. 53. The drag coefficient of very long cylinders becomes more and more nearly equal to that of the infinitely long one where two-dimensional flow takes place.

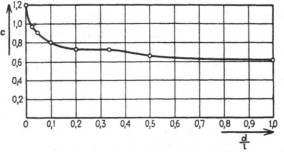

Fig. 53.—Drag coefficient *vs.* slenderness of cylinder d/l for $R = wd/\nu = 10^5$.

Figure 54 shows the resistance curve for a sphere and also for a circular disk perpendicular to the flow. The individual experimental points have not been plotted in the curve, but it has been found that the law of similarity holds well so

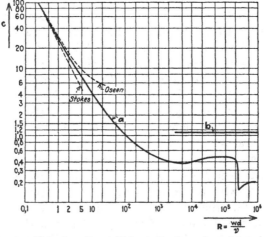

Fig. 54.—Drag coefficient *vs.* Reynolds' number for sphere (*a*) and circular disk (*b*) according to Wieselsberger.

that the individual points for spheres and disks of various diameters and at various velocities lie on smooth curves. For comparison, the theoretical laws of Stokes and Oseen for small Reynolds' numbers are also drawn in the figure.

Figure 55 shows the curves $c = f(R)$ for a number of bodies of revolution of various degrees of slenderness. The two upper curves in the diagram are for an ellipsoid of rotation with an axis ratio of 1: 0.75 (the small axis parallel to the flow); the next curves designated by a are for a sphere. The next two pairs of curves are for ellipsoids of rotation with an axis ratio of 1:1.33 and 1:1.8 respectively (the small axis perpendicular to the flow). The curves designated by b give the drag coefficients for an airship model. It is seen that the transition between the high drag coefficient below the critical number to the small

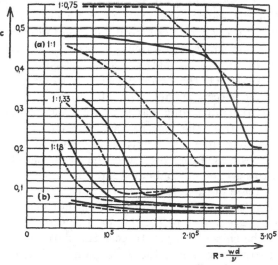

Fig. 55.—$c = f(R)$ for rotationally symmetrical bodies of various slenderness. Full lines for uniform wind; dashed lines for turbulent wind.

coefficient above the critical number is less definite and abrupt for slender bodies and for turbulent air than for blunt bodies and smooth air.

65. Resistance in Fluids with Free Surfaces; Wave Resistance.—In case the body is not entirely submerged in one fluid but moves through the surface of contact between two fluids of different density, a new kind of drag appears owing to the formation of waves. This is of great practical importance for the resistance of ships which are partially immersed in water and partially in air. The part of the total drag due to the air is generally so small that it can be neglected (except, of course, for sailing vessels).

The total resistance is the sum of three quantities: (1) the friction drag of the water on the hull of the ship, (2) the pressure drag of the water, and (3) the wave resistance. This last resistance is due to the fact that pressure differences at the ship body cause differences in the level of the water, which leave the ship in the form of waves. These waves carry a certain amount of kinetic energy with them away from the ship. The problem of the calculation of the wave resistance is intimately tied up with the amount of wave energy which is flowing per unit of time through a surface traced around the ship. The velocity with which the wave energy is leaving the ship is not the "phase velocity" of the waves, but rather their "group velocity," *i.e.*, the velocity of propagation of a group of waves in front of which, as well as behind which, the water is at rest. In a deep ocean of infinite extent (unlike a canal) the ship is accompanied by two systems of waves: (1) the "cross waves" of which the crests are approximately perpendicular to the direction of motion of the ship and (2) the "diverging waves" with about 40 deg. central angle emerging from the bow as well as from the stern of the ship. Depending upon the length of the ship and its velocity, the diverging waves of the bow and of the stern can interfere with each other to a greater or smaller extent. In case the interference causes a decrease in the intensity of the wave, the ship resistance becomes smaller, and conversely. The group velocity of these wave systems is equal to half their phase velocity, which in turn is equal to the ship velocity. In case the ship starts from rest, the wave system covers half the distance moved through.

These conditions are modified when a ship moves in a canal of a finite width or in shallow water. In the first case the system of diverging waves loses its importance; only cross waves appear of which the group velocity depends very much on the ratio of the ship velocity to the velocity of the free waves in the canal. For a ship velocity which is equal to or greater than a certain critical velocity, the group velocity becomes equal to the phase velocity, *i.e.*, the wave energy moves with the ship and consequently is not dissipated. This leads to a very small wave resistance. However, the amplitude of the waves is also a function of the ship velocity, and it happens that this amplitude becomes a maximum at the critical velocity mentioned above. The combined effect is that the wave resistance increases to a maximum when the ship velocity approaches the critical velocity, and then suddenly drops practically to zero when the critical velocity is exceeded.[1]

66. The General Resistance Law.—Since ship waves are formed under the simultaneous action of inertia forces and gravity forces, mechanical similarity of the wave motions is obtained for geometrically similar ships if the Froude's numbers $F = V^2/lg$ are equal (V is the velocity of the ship, l its length, and g the acceleration of gravity), (see Art. 5). This relation was derived by completely neglecting the viscosity forces. The practical consequence of this law is that the wave system belonging to a certain ship model is similar to the wave system of the large ship itself if the velocities are proportional to the square roots of the lengths. If the wave resistance is assumed to be proportional to the projected area A, the density ρ and the square of the velocity V, the proportionality factor or wave-drag coefficient is equal for geometrically similar ships only if the Froude's numbers are equal. In general, therefore, the wave-drag coefficient is a function of Froude's number. In the light of the relation found in Art. 59, it can be stated that the drag coefficient of a body in a viscous, incompressible fluid with free surfaces is a function of Reynolds' number and of Froude's number:

$$c = f\left(\frac{wl}{\nu}, \frac{w}{\sqrt{lg}}\right).$$

For the sake of completeness it is mentioned that, if the compressibility of the fluid cannot be neglected, the third dimensionless parameter on which the drag depends is the ratio between the velocity w and the velocity of sound w_s in the undisturbed fluid:

$$w_s = \sqrt{\frac{dp}{d\rho}}.$$

Generally, if inertia, viscosity, gravity, and compressibility all influence the resistance, the expression for it can be written as[2]

$$D = A\frac{\rho w^2}{2}f\left(\frac{wl}{\nu}, \frac{w}{\sqrt{lg}}, \frac{w}{w_s}\right).$$

---

[1] MÜLLER, C. H., Hydrodynamics of the Ship (German), "Encyclopaedie der mathematischen Wissenschaften," vol. IV, No. 3, p. 563.

[2] Cavitation, capillarity, and heat conduction have not been considered in this formula.

In case *two* dimensionless quantities are essential for the drag, mechanical similarity is not possible when using the same fluid, since two of these dimensionless variables cannot simultaneously remain constant when l is changed.

In Chap. XIII, "Fundamentals,"[1] it is shown that in all cases where the ratio of the velocity of fluid to the acoustic velocity is small with respect to unity, the compressibility can be neglected.

In the case of ship resistance, the friction drag or skin friction is less important than the pressure and wave resistances, since it depends only slightly on the shape of the ship. Therefore Froude's similarity law is usually applied to the model tests. The friction drag is determined by a separate experiment and is then subtracted from the total resistance measured at the model. The "residual" thus obtained is converted to the large ship by means of Froude's rule and to this residual the friction drag of the ship is added. This procedure we owe to Froude and it is applied generally to model resistance experiments. Recently, Telfer[2] has proposed a somewhat different procedure.

67. Resistance to Potential Flow.—It is a deplorable fact that no theory of drag yet exists which even approximately does justice to the experimental results. A relatively unimportant exception is the case of "creeping" motion, $R \ll 1$, which will be discussed in Art. 73. The general differential equations of viscous fluids lead to mathematical difficulties which may be unconquerable for a long time to come.

Therefore an attempt was made to solve first the very much simpler problem of the motion of solid bodies through the ideal fluid (incompressible, homogeneous, and without viscosity). The integration of the differential equations for this case has been accomplished with much ingenuity and often with great mathematical complication. However it was found that it is not permissible to neglect the viscosity completely, even in case it becomes infinitely small (see Arts. 1 and 44). A great amount of literature exists on the theoretical solution of the flow around moving bodies in the ideal fluid. However we shall not dwell on these theories here, since they do not lead to a solution which conforms in the least with actually occurring flow phenomena.

---

[1] See footnote, p. 3.

[2] TELFER, W., "Frictional Resistance and Ship Resistance Similarity," paper read before the N.E. Coast Institute of Engineers and Ship Builders, November, 1928 (London, 1929).

68. Drag of a Sphere Is Zero for Uniform Potential Flow.—
Only one special case will be discussed, since its mathematical
solution is relatively simple—the motion of a sphere. Utilizing
the method of sources and sinks, the potential function of a
sphere of radius r_0 moving through
a liquid at rest with the velocity a
is found to be

$$\Phi = \frac{a}{2}\frac{r_0^3}{r^2}\cos\varphi$$

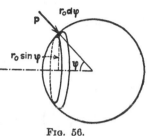

Fig. 56.

(see Art. 70, "Fundamentals"[1]).
Since for a fluid without viscosity
an action on the body can take place
only in the form of pressure on its surface, the expression for
the drag, *i.e.*, for the force in the direction of motion, becomes
(Fig. 56)

$$D = \int_0^{2\pi} 2\pi r_0 \sin\varphi \cdot r_0 d\varphi \cdot p\cos\varphi.$$

The general case of a non-steady flow will be considered. The
pressure p has to be computed from the general equation of
Bernoulli, which can be written as

$$\frac{\partial\Phi}{\partial t} + \frac{w^2}{2} + \frac{p}{\rho} = \text{const.},$$

(see page 127, "Fundamentals"[1]) assuming constant density and
the action of gravity to be neutralized by buoyancy. The
constant on the right-hand side depends on the time, since an
infinite ocean of the fluid is assumed in which the pressure at
great distances from the sphere remains constant.

The expression $\partial\Phi/\partial t$ corresponds to a definite point in steady
space, while r has to be taken from a moving point. Choosing
the x-direction such that it coincides with the direction of the
velocity a, the rate of change with the time of the velocity
potential Φ expressed in terms of a coordinate system at rest with
respect to the fluid at infinity becomes

$$\frac{\partial\Phi}{\partial t^\star} + a\frac{\partial\Phi}{\partial x}.$$

[1] See footnote, p. 3.

Here ∂t^* means a differentiation in a system moving with the sphere. Substituting the relation $x = r \cos \varphi$, this becomes

$$\frac{\partial \Phi}{\partial t} + a\left(\frac{\partial \Phi}{\partial r} \cos \varphi - \frac{\partial \Phi}{r\partial \varphi} \sin \varphi\right).$$

Performing the differentiations, we get

$$\frac{1}{2}\frac{r_0^3}{r^2} \cdot \frac{\partial a}{\partial t} \cos \varphi - a^2\frac{r_0^3}{r^3}\left(\cos^2 \varphi - \frac{1}{2} \sin^2 \varphi\right).$$

Observing that

$$w^2 = \left(\frac{\partial \Phi}{\partial r}\right)^2 + \left(\frac{\partial \Phi}{r\partial \varphi}\right)^2 = a^2\frac{r_0^6}{r^6}\left(\cos^2 \varphi + \frac{1}{4} \sin^2 \varphi\right),$$

we therefore obtain for p/ρ:

$$\frac{p}{\rho} - \text{const.} = -\frac{1}{2}\frac{r_0^3}{r^2} \cdot \frac{\partial a}{\partial t} \cos \varphi + a^2\frac{r_0^3}{r^3}\left(\cos^2 \varphi - \frac{1}{2} \sin^2 \varphi\right),$$

$$-a^2\frac{r_0^6}{r^6}\left(\cos^2 \varphi + \frac{1}{4} \sin^2 \varphi\right).$$

The constant in this equation is the pressure at infinity p_0 divided by the density ρ, which can be seen by letting r approach infinity.

For points on the surface of the sphere, *i.e.*, for $r = r_0$, we get after some transformations

$$\frac{p}{\rho} = -\frac{1}{2}r_0\frac{\partial a}{\partial t} \cos \varphi + \frac{9}{16}a^2 \cos 2\varphi - \frac{a^2}{16} + \frac{p_0}{\rho}.$$

Substituting this value for p in the formula for the drag, it is found that the integrals of the last three terms become identically equal to zero, so that the total result is

$$D = \frac{2}{3}\pi\rho r_0^3\frac{\partial a}{\partial t}.$$

In case the sphere moves with constant velocity, *i.e.*, in case $\partial a/\partial t = 0$, it is seen that no drag occurs. It is interesting to note that the individual fluid particles which are pushed aside by the sphere in its motion do not return to their former positions. The paths of the individual particles are not closed curves but of the shape shown in Fig. 57.[1] Besides pushing the particles aside temporarily in passing, the sphere also displaces the fluid

[1] RIECKE, E., Notes on Hydrodynamics (German), *Nachr. Ges. Wiss. Göttingen, Math.-phys. Klasse*, p. 347, 1888.

particles permanently in the direction of its motion. This displacement, however, is of importance only in the vicinity of the wake.

69. Resistance Due to Acceleration.—For an accelerated motion, it was seen that potential flow does lead to a resistance. In order to accelerate a sphere in an ideal fluid it is not only necessary to exert a force equal to the product of the mass of the sphere and its acceleration, but an additional force

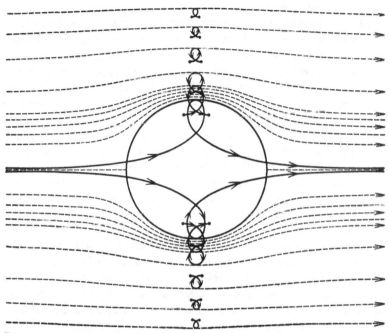

Fig. 57.—Absolute paths of particles for rectilinear motion of sphere through ideal fluid. (*E. Riecke.*)

is required to accelerate the mass of the fluid particles set in motion by it. From the above equation for the resistance, it is seen that this additional force is equal to the product of the acceleration of the sphere and the mass of an amount of fluid of half its volume. The motion of a sphere in an infinite ocean of an ideal fluid is therefore completely identical with the motion of the sphere in a vacuum if its mass is increased with that of an amount of liquid equal to half its volume.

The apparent increase in mass for various bodies depends on the shape and on the direction of motion. For instance, for the two-dimensional flow round a circular cylinder, the apparent increase in mass is equal to the full mass of the cylinder in liquid. For other cylinders of non-circular cross section the apparent increase in mass can be calculated also.

These considerations for spheres and circular cylinders have no great practical importance since the actual flow is entirely different from the theoretical one.[1] However, for bodies of airship shape, where the actual flow is very much similar to the calculated potential flow, these investigations have some value.

70. Application of the Momentum Theorem.—The theorem of no resistance of a sphere in uniform motion through ideal fluid, often referred to as the "paradox of Dirichlet," can be easily extended to bodies of arbitrary shape by means of the momentum theorem. If the body in its uniform motion would have a drag, the fluid should show an increased momentum. This should be detectable by integration on any closed surface traced around the body (see Art. 100, "Fundamentals"[2]). Assuming a motion starting from rest in an infinite ocean of fluid, the velocity potential of the flow at sufficiently large distances from the body decreases with $1/r^2$. Consequently, the velocity decreases with $1/r^3$. Since, from then on, the body is supposed to move at a constant velocity, the pressure p, being of the order $\rho a w$, also changes with $1/r^3$.

The flow phenomenon considered before is a non-steady one. It can be made steady by giving the fluid and the body a velocity opposite and equal to the velocity of the body, which puts the body to rest. Around it we trace a sphere C_1 and proceed to determine the momentum flowing through this sphere as well as the pressure integral over it (see Art. 100, "Fundamentals"[2]):

$$\oint^{C_1} d\mathbf{S} \circ \mathbf{w}_1 \mathbf{w}_1 + \oint^{C_1} p_1 d\mathbf{S} = \mathbf{D}.$$

Writing down the same expression for other spheres C_2, C_3, etc., with radii r_2, r_3, etc., it is found that both integrals decrease steadily with increasing distance from the body, owing to the fact that the velocities and the pressures decrease with $1/r^3$ while the areas of the spheres increase only with r^2. For the

---

[1] For small oscillations (where the amplitude is small with respect to the radius), the potential theory is found to agree with experiment.

[2] See footnote, p. 3.

infinite radius both integrals vanish. Since their sum is independent of r, it follows that it must be equal to zero. This proves the theorem that a body of arbitrary shape moving uniformly through an infinite ocean of ideal fluid does not experience any drag.

It is noted that the proof supposes an infinite extent of fluid and only a single body, since otherwise it is not possible to proceed to the limit $r = \infty$. It will be discussed later that, if there is more than one body or if there are a single body and a wall, forces between the bodies do occur even in the case of a fluid without viscosity. However, according to the energy theorem, these forces cannot be of the nature of a drag since the

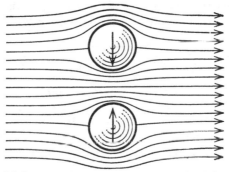

FIG. 58.—Potential flow round two spheres. Since velocity between is great, the spheres attract each other.

energy is not dissipated but remains in the neighborhood of the body. The only exception to this rule is if free surfaces exist so that energy can be dissipated by means of waves traveling away from the body (see Art. 65).

71. Mutual Forces between Several Bodies Moving through a Fluid.—When more than one body is moving through an ideal fluid, a mutual force is exerted between them, which, however, is generally very small. As an example, consider the potential flow round two spheres of which the line joining the centers is perpendicular to the direction of the flow (Fig. 58). The streamlines between the two spheres are crowded together more than the lines outside, which, according to Bernoulli's equation, implies a smaller pressure between the spheres. Consequently, owing to the relatively greater pressure on the outside, the two spheres are pushed towards each other, which creates the impression that they are attracting each other.

Another example is when two spheres are situated one behind the other so that the line joining the centers is in the direction of the flow (Fig. 59). In this case the streamlines between the two spheres are spaced farther apart than those outside, involving a larger pressure here, which leads to an apparent repelling. The forces involved in this effect are very small; they are inversely proportional to the fourth power of the mutual distance. If the line of symmetry in Fig. 58 is replaced by a solid wall of infinite dimensions, the picture of the flow is apparently not influenced, so that a sphere moving in an ideal fluid parallel to a plane wall is attracted by that wall.

72. Resistance with Discontinuous Potential Flow.—The great discrepancy between the theory of potential flow and the experimental observations soon caused endeavors to modify

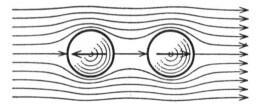

Fig. 59.—Potential flow round two spheres. They repel each other.

the theory to such an extent as to calculate at least some resistance. These early attempts did not strike at the root of the trouble, *i.e.*, at the viscosity of the fluid. The difficulties involved in the integration of the general equation of the viscous fluid seemed unsurmountable, and they still are. The attempts rather consisted of assuming certain discontinuities in the velocity field, such, for instance, as observed in the formation of a jet coming out of an orifice.

The following discussion serves as a justification for the assumption of surfaces of discontinuity, in which the velocity changes suddenly (see Art. 92, "Fundamentals"[1]). Consider a circular cylinder of infinite length moving with the uniform velocity u perpendicularly to its axis (two-dimensional flow). There are two points P_1 and P_2 on the circle in the middle between the two points of stagnation of the flow, in which the tangential velocity reaches its maximum value $2u$. In order to prevent the absolute pressure at those points from becoming negative, which

[1] See footnote, p. 3.

apparently is physically impossible, it is necessary that the pressure at infinity be at least $\frac{3}{2}\rho u^2$.[1] Considering instead of the circular cylinder one of elliptical cross section with the major axis perpendicular to the direction of flow, the two points P_1 and P_2 of maximum velocity show velocities greater than for the circular cylinder. This velocity increases for a diminishing minor axis of the elliptic cylinder, and in case the minor axis becomes zero, and the cylinder reduces to a line perpendicular to the direction of flow, the velocities at its edges would become infinite. Therefore it is seen that for an elliptic cylinder of a decreasing minor axis the necessary pressure at infinity becomes larger and larger, and in the case of a flat plate it has to become infinitely large in order to prevent negative pressures at these points.

These difficulties, caused by the fact that solutions of continuous potential flow in certain cases cannot be made to satisfy the physical requirements regarding pressure, were first avoided by Helmholtz[2] by assuming surfaces of separation across which the tangential velocity experiences a sudden change. This assumption of surfaces of discontinuity seems to be justified by experiment in so far as in the actual case no flow around the sharp edges of the plate is observed. The fluid rather breaks away from these edges, thus causing a region of dead water behind the body. A good explanation of the creation of surfaces of discontinuity, however, can be given only on the basis of the theory of the boundary layer. That such surfaces are not observed in practice is due to the fact that they are unstable against small disturbances.

The method of discontinuous surfaces represents real progress as compared to the theory of continuous potential flow since the new theory leads to a calculated resistance proportional to the

---

[1] Let p_0 be the pressure at an infinite distance from the body, where the undisturbed velocity is equal to u. Let p' be the pressure at the two points of maximum velocity $2u$. Bernoulli's equation then reads

$$\frac{u^2}{2} + \frac{p_0}{\rho} = \frac{(2u)^2}{2} + \frac{p'}{\rho}$$

so that

$$p' = p_0 - \frac{3}{2}\rho u^2.$$

[2] HELMHOLTZ, H., On Discontinuous Fluid Motions (German), Monatsber. Kgl. Akad. Wiss. Berlin, 1868, p. 215, or Two Hydrodynamical Essays (German), *Ostwalds Klassiker*, No. 79.

projected area, the density, and the square of the velocity, which is in accordance with the experimental results.   The actual calculation carried out by Kirchhoff[1] (see Art. 82, "Fundamentals"[2]) for the case of a plate perpendicular to the flow led to a drag coefficient which is far too small.   The calculated value for

c is $\dfrac{2\pi}{4 + \pi}$ = 0.88, while the experiment gives c = 2.0.

The great discrepancy between these two results is due to the fact that in the actual case there is a partial vacuum in the dead-water region, which cannot be taken account of by the method of Kirchhoff. Moreover, the calculated streamline picture behind the plate is considerably different from the experimental one.   The theoretical surfaces of discontinuity extend to infinity approximately like two parabolic arcs with their apices somewhat displaced (Fig. 60).   In the actual case, however, the flow closes together somewhat at some distance behind the plate and then is mixed up with the irregular eddies. Owing to the internal friction in the fluid, these irregularities in the velocity are damped down more and more so that at a great distance behind the plate there is approximately undisturbed flow.

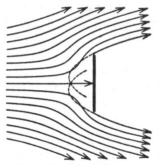

Fig. 60.—Discontinuous potential flow round plate (two dimensional).   The boundary consists approximately of two parabolic arcs.

A serious limitation of the method of discontinuous potential flow is that it can be applied practically only to two-dimensional motions.   In the three-dimensional case, where the method of complex functions has to be replaced by the general theory of potential flow, the difficulties encountered are very much greater even in the relatively simple case of rotational symmetry. Extensive literature on this subject is quoted by Jaffé.[3]

73. Stokes's Law of Resistance.—For very small Reynolds' numbers, where the inertia forces become small with respect to

[1] KIRCHHOFF, G., The Theory of Free Fluid Jets (German), *Crelles J.*, vol. 70, 1869.

[2] See footnote, p. 3.

[3] JAFFÉ, Discontinuous and Multivalued Solutions of the Hydrodynamical Equations (German), *Z. angew. Math. Mech.*, vol. 1, p. 398, 1921.

the viscosity forces, the general differential equation of Navier-Stokes can be integrated by completely neglecting the inertia forces. In the case of these so-called "creeping motions," the general differential equation

$$\frac{D\mathbf{w}}{dt} = -\frac{1}{\rho} \operatorname{grad} p + \nu\Delta\mathbf{w},$$

by neglecting the inertia member, reduces to

$$\mu\Delta\mathbf{w} = \operatorname{grad} p.$$

This equation in combination with the equation of continuity

$$\operatorname{div} \mathbf{w} = 0$$

was first solved by Stokes[1] for the case of the sphere, considering the complete boundary condition of no tangential velocity of the fluid along the surface of the obstacle. With the aid of a stream function invented by him, he found for the drag **D** of a sphere of radius r moving with a velocity **w** through an incompressible viscous fluid of infinite extent the expression

$$\mathbf{D} = 6\pi\mu r\mathbf{w}.$$

Kirchhoff[2] has given a simplified derivation of this formula.

Assuming the constant force of gravity to be acting on the sphere, *i.e.*, considering its falling motion in a very viscous fluid, the velocity of the sphere evidently will be constant as soon as the resistance has become equal to the weight of the sphere in the surrounding liquid. Let ρ_1 be the density of the sphere and ρ the density of the liquid; then this condition is reached when

$$\mathbf{D} = 6\pi\mu r\mathbf{w} = \tfrac{4}{3}\pi r^3(\rho_1 - \rho)\mathbf{g}.$$

The velocity of descent therefore is

$$\mathbf{w} = \frac{2}{9}r^2\mathbf{g}\frac{\rho_1 - \rho}{\mu}$$

In the case of small water particles in air this becomes numerically

$$w = 1.3 \ 10^6 r^2 \qquad \text{(c g s units)}.$$

[1] STOKES, G., *Trans. Cambridge Phil. Soc.*, vol. 8, 1845 and vol. 9, 1851; or *Collected Papers*, vol. 1, p. 75.

[2] KIRCHHOFF, G., "Lectures on Mathematical Physics," (German), 4th ed. vol. 1, p. 378, 1897.

Since this law of Stokes is an approximate one for very small Reynolds' numbers only (R smaller than about $wr/\nu = 0.5$), it follows that there is an upper limit for the radii of falling spheres above which the flow cannot be represented any more by this equation. For small drops of water falling in air, creeping motion occurs for radii smaller than 0.002 in. corresponding to a Reynolds' number of approximately $R = 0.50$. It is seen therefore that Stokes's solution of the equations of hydrodynamics is applicable to the fall of water particles in clouds or fog but not to the falling motion of rain drops. It is also applicable to artificial fogs like those created in the experiment of Thomson and Wilson for determining the charge of an electron.

74. Experimental Verification for Water; Influence of the Walls of the Vessel.—Experimental verifications of the law of Stokes have been given by several investigators. We mention especially the work of Allen, Ladenburg, Arnold, and Zeleny-McKeehan. The experiments of Liebster[1] are applicable primarily to greater Reynolds' numbers (R between 0.2 and 500), which range has been investigated also by Arnold and Allen.

Allen[2] inserted very small bubbles of air into water ($\mu = 0.012$ g/cm. sec; 12°C) or in anilin ($\mu = 0.006$; 12°C) by means of very fine capillary tubes of glass. He compared the experimental velocities of these bubbles with those calculated from Stokes's formula. In order to determine whether gas bubbles behave in the same manner as solid spheres he also investigated the velocities of small spheres of paraffin in water as well as those of amber in anilin. Figure 61 shows the experimental results recalculated to dimensionless quantities; the drag coefficient

$$c = \frac{D}{r^2\pi\frac{\rho}{2}w^2} = \frac{\frac{4}{3}r^3\pi(\rho_1 - \rho_2)g}{r^2\pi\frac{\rho}{2}w^2}$$

as a function of $R = wd/\nu$.

Ladenburg[3] made his experiments with steel spheres in "Venetian turpentine," a mixture of turpentine and rosin ($\mu = 1,300$

---

[1] LIEBSTER, H., On the Resistance of Spheres (German), *Ann. Physik*, vol. 82, p. 541, 1927.

[2] ALLEN, H. S., The Motion of a Sphere in a Viscous Fluid, *Phil. Mag.*, vol. 50, p. 323, 1900.

[3] LADENBURG, R., On the Viscosity of Fluids and Its Relation with Pressure (German), *Ann. Physik*, vol. 22, p. 287, 1907.

g/cm. sec; 16°C). He noticed the considerable influence of the walls of the vessel on the resistance of the spheres. He found, for instance, that the drag of a sphere of 3-mm diameter in a vessel of 27-mm diameter was 15 per cent greater than the resistance of the same sphere in a vessel of 44-mm diameter ($R = 10^{-5}$). He states that even when the diameter of the vessel is 90 times greater than the diameter of the sphere, an increase

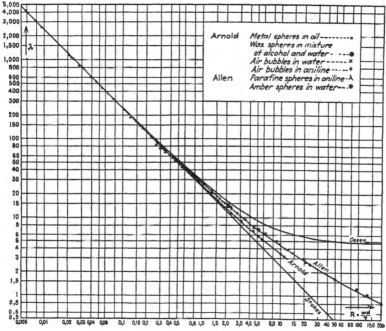

FIG. 61.—Drag coefficient *vs.* Reynolds' number for sphere according to tests of Arnold and Allen and according to theories of Stokes and Oseen.

in the drag due to the influence of the walls of the vessel can be detected. In all cases, the viscosities calculated from his measurements by means of Stokes's law are greater than those calculated by the law of Hagen-Poisseuille from experiments of the flow through tubes (Art. 20). Therefore the method of falling spheres cannot be used indiscriminately for the measurement of the viscosity of fluids.

The method of theoretical calculation of the influence of the walls, initiated by Lorentz,[1] was carried out by Ladenburg

---

[1] LORENTZ, H. A., "Lectures on Theoretical Physics" (German), vol. I, p. 23, Leipzig, 1907.

for the case of an infinitely long tube (see also Weyssenhoff[1]). Applying the correction as found by this theory, the experimental values of μ found by means of Stokes's formula agree among themselves within 1 per cent; however, they are still about 3 per cent greater than those found by the experiment of flow through tubes. A probable explanation of this discrepancy Ladenburg believes to be the influence of the cover and the bottom of the vessel.

Arnold also determined the viscosity by means of the fall method and Stokes's law. He investigated from which sphere diameter up the viscosity thus determined differs from the one found by the method of Hagen-Poisseuille. His experiments were conducted in glass tubes so that it becomes necessary to correct the results for the influence of the walls. This correction applied by means of Ladenburg's formula leads to very good results. The spheres used were made of various metals of low melting point falling in some vegetable oil of which the temperature was kept constant within 0.1°C. Figure 61 shows the relation between the drag coefficient and the Reynolds' number calculated from these experiments.

75. Experimental Verification for Gases.—The experimental investigations discussed thus far are limited to liquids. Zeleny and McKeehan[2] investigated the applicability of the law to air. They used very small spheres made of wax, paraffin, and mercury, of diameters as small as 10^{-4} in., made by means of an atomizing procedure. These spheres were dropped in a tube of about 12-in. length and about ¼-in. diameter. The velocity of fall was measured by a special test procedure and agreed with the result of Stokes's formula within ½ per cent on the average. On the other hand, similar experiments by the same authors with spores of certain plants of microscopical dimensions (lycopodium $6 \cdot 10^{-4}$ in. diameter; locoperdon $8 \cdot 10^{-5}$ in. diameter; polytrichum $2 \cdot 10^{-4}$ in. diameter) gave velocities about 30 per cent smaller than those calculated by Stokes's formula.

The Reynolds' number below which a creeping motion takes place and above which the effects of inertia become of importance is about 0.2 to 0.5. Below this limit, therefore, the drag is

[1] WEYSSENHOFF, J., Investigations on the Validity of the Formula of Stokes-Cunningham (German), *Ann. Physik*, (4), vol. 62, p. 1, 1920.

[2] ZELENY, J., and L. W. McKEEHAN, The Falling Velocity of Small Spheres in Air (German), *Physik. Z.*, vol. 11, p. 78, 1910.

détermined sufficiently accurately by Stokes's formula. In case, however, that the dimensions of the obstacles become so small as to be comparable to the mean free path of the molecules of the liquid or the gas, Stokes's law ceases to be valid, since it was derived on the assumption of a continuous medium. Cunningham[1] has made a theoretical, and Millikan[2] an experimental investigation of the change in Stokes's formula, when the assumption of a continuous medium is dropped. This lower limit of Stokes's law is very small in gases of common pressures and in liquids. For instance, Perrin[3] found the law to be valid for spheres of $4 \cdot 10^{-6}$ in. diameter in air. A critical survey of this subject with an extensive quotation of the literature is given by E. Meyer and W. Gerlach.[4]

76. Correction of Stokes's Law by Oseen.—The characteristics of the flow in the vicinity of the sphere are approximated very well by Stokes's theory and consequently the drag is approximated very well also. Oseen,[5] however, has shown that at large distances from the body the assumption that the inertia forces are negligible with respect to the frictional forces does not hold. This is seen from a comparison of the order of magnitude of the inertia members in the equation, for instance, $\rho u\, du/dx$ with that of the frictional members, for instance, $\mu \Delta u$ (see page 8). In Stokes's solution the velocity at great distances from the sphere can be considered to be equal to the velocity of the body w diminished by an amount proportional to wl/r, where l is a characteristic length of the body and r the distance away from it. It is seen that the inertia forces, proportional to $\rho w^2 l/r^2$, become large with respect to the frictional forces, proportional to $\mu wl/r^3$. At large distances, therefore, the assumption of the theory of Stokes is by no means satisfied. It has to be noted, however, that, although the ratio between the inertia forces and the viscosity forces increases at greater distances from the sphere, these forces themselves decrease with $1/r^2$ and

---

[1] CUNNINGHAM, *Proc. Roy. Soc. (London)* (A), vol. 83, p. 357, 1910.

[2] MILLIKAN, *Phys. Rev.*, April, 1911.

[3] PERRIN, J., The Law of Stokes and the Brownian Movement (French), *Compt. rend.*, vol. 147, p. 475, 1908.

[4] MEYER, E., and W. GERLACH, On the Validity of Stokes' Formula and the Mass Determination of Ultra Microscopic Particles (German), *Festschrift für Elster u. Geitel*, Brunswick, 1915.

[5] OSEEN, C. W., On Stokes' Formula (German), *Arkiv Mat. Astron. Fysik*, vol. 6, 1910; vol. 7, 1911.

$1/r^3$ respectively.   Therefore, the correction as applied by Oseen becomes of importance only in that part of the field where the velocities have become so small that they have ceased to be of great influence.   Because of this, the interest of Oseen's correction is mainly theoretical.   The experimental results available today are not sufficient to determine whether Oseen's formula or Stokes's formula gives better results for Reynolds' numbers of the order of magnitude 1.   For the case of a circular cylinder, Stokes's method of calculation did not lead to any result at all, and the new theory of Oseen was necessary to obtain one.   The calculations involved in this process were carried out by Lamb.[1]

77. The Resistance of Bodies in Fluids of Very Small Viscosity.—It was seen that the formula of Stokes and the correction of Oseen apply only to creeping motions, *i.e.*, motions for very small Reynolds' numbers.   In case the inertia forces in the fluid become of the same order of magnitude as the viscosity forces, the theory breaks down.   The discrepancy becomes very large when with still greater Reynolds' numbers the fluid breaks away from the body at certain points and executes an eddying motion apparently without any regularity.

In all cases of this kind where it seems impossible to obtain results by means of direct theoretical calculations, it is often useful to apply momentum and energy theorems in order to find at least approximate results.   As an introduction to this method of attack two theorems will be discussed: (1) one dealing with the resistance of the "half body" and (2) one with the momentum of a source.

78. The Resistance of the Half Body.—By a half body we mean a body of which one end is situated in the field of flow while its other end extends to infinity.   The relations which can be found for such a half body apply approximately to the front part of an airship, since the tail of the ship is so far away that it hardly affects the flow round the front end.   Besides being useful for this practical application, the theorems of the half body will prove to be of fundamental interest also.   The fluid is assumed to be completely without friction and we want to determine the drag of this half body in a potential flow.   However, the problem is indeterminate as long as no definite statement is made regarding the pressure at the rear end of the body.

[1] LAMB, H., On the Uniform Motion of a Sphere through a Viscous Fluid, *Phil. Mag.*, vol. 21, p. 120, 1911.

In order to circumvent this difficulty it is assumed that at a sufficient distance from the front end there is a slit into which the surrounding pressure penetrates (Fig. 62). The pressure drag of the half body is then understood to be the resulting force due to the pressure differences on the part thus cut off.

By means of the method of sources and sinks, as discussed in Art. 69, "Fundamentals,"[1] it is possible to calculate the flow round the half body, and therefore the pressure drag can be calculated by integrating the momentum and the pressure over a sufficiently large bounding surface.

The result, however, can be obtained in a simpler manner by the use of a device, namely, by considering the resistance of a

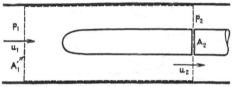

Fig. 62.—Flow round half body in cylinder.

half body in a wide hollow cylinder, while the absence of friction is still assumed (Fig. 62). The surface of integration is represented by the dotted line. Letting the ratio of the cross section A_2, of the half body to that of the cylinder A_1 be $\alpha = A_2/A_1$, the equation of continuity gives

$$A_1 u_1 = (A_1 - A_2)u_2$$

or

$$u_1 = (1 - \alpha)u_2.$$

By means of Bernoulli's equation the last result can be written as

$$p_1 - p_2 = \frac{\rho u_2^2}{2}[1 - (1 - \alpha)^2].$$

The momentum theorem applied to the part of the body to the left of the slit (in which the pressure equals p_2) leads to a resistance

$$D = A_1(p_1 - p_2) + A_1 \rho u_1^2 - (A_1 - A_2)\rho u_2^2$$

or, applying the continuity equation,

$$D = A_1(p_1 - p_2) + A_1 \rho u_1(u_1 - u_2).$$

[1] See footnote, p. 3.

Substituting in this the result for $p_1 - p_2$ obtained before,

$$D = A_1\frac{\rho u_2^2}{2}[1 - (1 - \alpha)^2 + 2\{(1 - \alpha)^2 - 1 + \alpha\}]$$

or

$$D = A_1\frac{\rho u_2^2}{2} \cdot \alpha^2 = A_2\frac{\rho u_2^2}{2} \cdot \alpha.$$

The drag coefficient referred to the velocity u_2 therefore becomes equal to $\alpha = A_2/A_1$. Letting the cross section A_1 of the tube

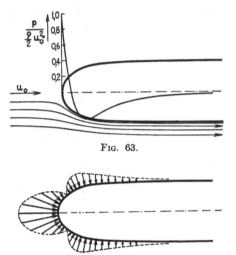

Fig. 63.

Fig. 64.

Figs. 63 and 64.—Pressure distribution on nose of blunt rotationally symmetrical body.

increase beyond all limits, it is seen that α converges to zero. In other words, the resistance of a half body in an infinite fluid is zero.

A physical explanation of this fact can be obtained by considering the pressure distribution at the front end of the half body, assuming that the flow does not break away. At the nose of the body it is seen that the streamlines at some distance away from it are convex with respect to its surface, which implies an excess pressure on the nose. Somewhat farther behind, however, the streamlines turn their concave side toward the body so that a diminished pressure exists there. Figure 63 shows the streamlines on the half body whereas Fig. 64 is a graphical representation of the pressure distribution. It is seen that the excess pressure and the sucking action are approximately in equilibrium.

In cases where behind a blunt obstacle there is a dead-water region or wake which becomes cylindrical at a great distance away from it, the assembly of the body and its wake form a half body and the flow round the obstacle plus the dead water differs in no respect from the flow round an ideal half body. It is concluded from this that a cylindrical wake leads to a zero resistance. Consequently, in order to get a drag, it is necessary for the wake to increase its cross section indefinitely. It was discussed in Art. 72 that Helmholtz and Kirchhoff investigated flow phenomena with dead-water regions increasing parabolically toward infinity, resulting in a definite drag.

79. Momentum of a Source.[1]—It has been shown before that the flow round a half body can be caused by a source or by a number of sources. Let u_0 be the undisturbed velocity and A the cross section of the cylindrical part of the half body. The total intensity of all the sources, *i.e.*, the volume of fluid generated per second is equal to $Q = Au_0$. If we consider instead of the half body (in which the source is merely existent in our imagination) an actual flow with a source, it is seen that an amount Q flows out more than in. Surrounding the source at a sufficient distance with a surface of integration, the influence of the source is felt as an additional flow of momentum $\rho Q u_0 = \rho A u_0^2$, leading to a negative resistance of that same numerical value. Since a half body does not experience any force it follows that a source in a uniform flow experiences a negative resistance of magnitude $\rho Q u_0$.

NOTE: The conditions can be understood still better by assuming the streamline surface passing through the point of stagnation to be a thin solid shell. The outer flow then represents the usual half-body flow round a solid shell. The inner flow consists of a source directing its flow toward infinity (Fig. 65). We know that the sum of the pressure forces existing on the outside and inside of the shell has to be zero, since the shell coincides with a free surface of streamlines and could be obliterated without changing the flow. On the other hand, the resultant pressure on the outside of the shell is zero according to the laws of the half body; consequently the resultant pressure on the inside must be zero also. It must be concluded that every second an amount of momentum $\rho Q u_0$ flows through any surface enclosing the source and lying completely inside the shell. We can then shrink this surface together indefinitely around the point source and still obtain the

[1] Articles 79, 80, and 81, are original contributions by L. Prandtl.

negative resistance $\rho Q u_0$ as a reaction. Actually in each point of space, and consequently also in the immediate vicinity of the source, the velocity consists of the sum of the source flow and the undisturbed flow u_0. The source flow by itself has no momentum because it extends symmetrically in all directions. The total flow, however, obtains momentum because of the fact that each particle at its birth receives a velocity u_0 which corresponds to a hypothetical driving force in the source. This hypothetical force finds its reaction in the negative resistance found before.

Now the theorem of the momentum of a source in a uniform flow will be proved in a different manner which will be of use in the subsequent considerations on the drag of a body. In principle, any kind of "bounding surface" is useful—a concentric sphere, two infinite planes before and behind the source, etc. In this case we choose a cylinder of which the axis is parallel

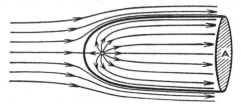

Fig. 65.—Half body with generating source.

to the flow at infinity. The cylinder is closed off by two faces far in front of, and far behind, the source. In order to calculate the force exerted on the cylinder in the direction of its axis, we have to calculate first the integral of the pressures and second the flow of momentum (Art. 100, "Fundamentals"[1]). The pressures on the curved surface of the cylinder do not contribute anything to the force in the desired direction since they are perpendicular to it. On the faces of the cylinder there is a pressure difference (excess pressure in front and vacuum behind). If, however, the two faces of the cylinder are moved to infinity, these pressure differences disappear (inversely proportional to the square of the distance from the source), so that this part of the pressure integral also becomes zero in the limit.[2] For the same reason the contribution of the momentum flow across the faces disappears in the limit. Therefore we have to consider

---

[1] See footnote, p. 3.

[2] It is understood that the contribution of the faces to the pressure integral would not disappear in case the cylinder were growing similar to itself. In that case the face surfaces would increase proportionally to the square of the distance, and the pressure integral therefore would tend to a finite limit.

only the momentum contribution on the curved surface of the cylinder. Owing to the presence of the source, some fluid flows through this surface to the outside. In Fig. 66 we consider two surface elements of the same size and situated symmetrically with respect to the source. Owing to the source alone, the velocities at these two elemental surfaces have symmetrical directions. However, besides these velocities there is the velocity of uniform flow u_0. Since the longitudinal velocity components due to the source alone are opposite in direction, the mean value of the longitudinal components of the complete velocities at the two elemental surfaces is equal to u_0. This relation holds for any two surface elements situated symmetrically with respect to the source. The total volume of fluid passing through the

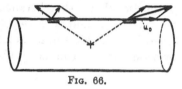

FIG. 66.

curved surface of the cylinder becomes equal to the intensity of the source Q if the cylinder is extended to infinity. Therefore the flow of momentum is calculated to be $\rho Q u_0$, which is the same result as obtained before.

The proof which has just been completed can be made the starting point for a derivation of the half-body theorem. It has to be considered that with the half-body flow the amount of fluid $Q = Au_0$ is lacking at the back end, since it is occupied by the solid body. Because of this effect, there is a shortage in the momentum of the amount $\rho Q u_0$, which is felt as a drag. This drag combined with the negative resistance of the source again leads to a total resistance equal to zero for the half body.

80. The Resistance of a Body Calculated from Momentum Considerations.—If a body in steady motion through a fluid at rest experiences a drag, it must be possible to prove the existence of this drag by means of the flow of momentum through a surface surrounding the body. In this respect, two different effects have to be considered.

1. There is a dead-water region or wake behind the body which has definite vortices at great Reynolds' numbers or no such vortices at small Reynolds' numbers. Owing to the action of viscosity or to the irregular vortex motion, the velocity in the wake becomes less and less at greater distances from the doby and the wake itself becomes wider. The wake conducts a

certain flow of momentum, the strength of which depends directly on the drag.

2. Further it is noted that the body together with its wake pushes the fluid away to the side so that it causes in the fluid outside the wake a flow apparently caused by a source. Figure 67 shows schematically the streamlines of such a flow for a coordinate system in which the undisturbed fluid remains at rest. It is possible to connect each streamline of the wake to a streamline of the source flow.

The width of the wake increases at a slower rate than is proportional to the distance from the body; the increase is between $\sqrt{x}$ and $\sqrt[3]{x}$ depending on the details of the flow. Therefore the kind of momentum consideration employed in the previous article can be applied here. According to the foot-

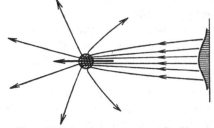

Fig. 67.—Apparent source and wake.

note on page 122 the pressure integral on the base surface of the cylinder becomes zero if its diameter increases at a slower rate than proportional to the distance. We assume that the wake region remains completely inside the curved surface of the cylinder. The flow in the wake becomes more and more nearly parallel at increasing distances from the body. Considering the pressure differences across the streamlines, it follows that far away from the body the pressure inside the wake is practically the same as in the source flow directly beside it.

After this introduction we proceed to apply the momentum theorem. As a system of reference we choose the coordinates in which the body is at rest, and the velocity of the fluid at infinity therefore is equal to u_0. The contribution of the curved surface of the cylinder is the momentum of the source, which is equal to a driving force $\rho Q u_0$ (where Q is the intensity of the source). The contribution of the two base surfaces of the cylinder is the difference between the flow of momentum in front of the obstacle, where the velocity is u_0, and behind, where the velocity is $u_0 - u'$, u' being the velocity in the wake relative to the body. Figure 68 shows the velocity distribution behind the obstacle with the characteristic "trough." Consequently the momentum flow becomes

$$\underbrace{\rho u_0^2 A}_{\text{goes in}} - \underbrace{\rho \int\int (u_0 - u')^2 dA}_{\text{goes out}} = 2\rho u_0 \int\int u' dA - \rho \int\int u'^2 dA.$$

In this expression, $\int\int u' dA$ is equal to the intensity of the source Q so that the first term reduces to $2\rho Q u_0$. The second term vanishes when proceeding to the limit of an infinite distance from the body, since it contains the wake velocity squared. Finally, taking together the contributions of the curved surface and the base surface, the total drag becomes

$$D = \rho Q u_0$$

so that

$$Q = \frac{D}{\rho u_0}.$$

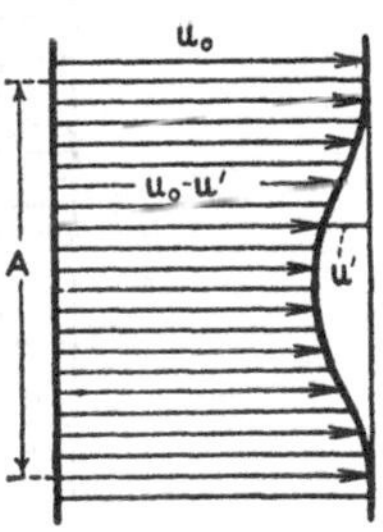

Fig. 68.—Velocities in wake.

Writing the drag in the usual form $D = cA\frac{\rho u_0^2}{2}$, we find for the intensity of the source

$$Q = \tfrac{1}{2}cA u_0,$$

which result is physically quite plausible. It is seen that the strength of the source is independent of the distance of the bounding surface from the body if this distance is sufficiently large. From this it is concluded that all streamlines in the wake extend from infinity directly to the body as indicated in Fig. 67.

If the motion of the body had started somewhere in the finite region, the wake would extend only to this point. The streamlines of the wake, which cannot end in any particular location, would then lead to a sink at the point of starting. This sink flow is accompanied by a vortex ring.

81. Method of Betz for the Determination of the Drag from Measurements in the Wake.—The relations found in Art. 80 are based on the velocity distribution in the wake at a great distance from the body. In making actual experiments, however, it is often necessary to come very much closer to the obstacle so that the analysis has to be extended to this more general case. This was done by Betz,[1] and the considerations of this article are based mainly on his publication.

As bounding surface for the integration two parallel planes are taken perpendicular to the direction of flow of the undisturbed fluid, one in front of and one behind the obstacle (Fig. 69). The velocities and the pressure in the front plane are u_1, v_1, w_1, p_1, and the corresponding ones in the rear plane are u_2, v_2, w_2, p_2. At

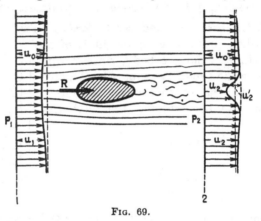

Fig. 69.

infinity we have $u = u_0$, $v = w = 0$, $p = p_0$. By means of the momentum theorem the drag becomes

$$D = \iint (p_1 + \rho u_1{}^2)dA - \iint (p_2 + \rho u_2{}^2)dA, \qquad (1)$$

in which the integrals extend over the entire area of both infinite planes. The problem consists of transforming these integrals in such a manner that the integration becomes restricted to the trough in the wake. To this end, we introduce the abbreviations

$$g_1 = p_1 + \frac{\rho}{2}(u_1{}^2 + v_1{}^2 + w_1{}^2),$$

$$g_2 = p_2 + \frac{\rho}{2}(u_2{}^2 + v_2{}^2 + w_2{}^2).$$

On any streamline which is not subjected to viscosity actions or to apparent friction due to turbulence, Bernoulli's theorem

[1] BETZ, A., A Method for the Direct Determination of Profile Drag (German), Z. Flugtechn. Motorluftschiffahrt, vol. 16, p. 42, 1925.

states that g = constant. Therefore $g_1 - g_2$ differs from zero only in the trough. For this reason, g_1 and g_2 are now substituted into Eq. (1) with the result that

$$D = \underbrace{\iint (g_1 - g_2)dA}_{\text{I}} + \underbrace{\frac{\rho}{2}\iint (u_1{}^2 - u_2{}^2)dA}_{\text{II}} +$$

$$\underbrace{\frac{\rho}{2}\iint \{(v_2{}^2 + w_2{}^2) - (v_1{}^2 + w_1{}^2)\}dA.}_{\text{III}} \quad (2)$$

The first integral in this expression already has the desired properties. In order to transform the second integral, we define a hypothetical flow which coincides with the actual flow everywhere except in the wake, where the hypothetical flow shall be such that $g_2' = g_1$ (there are no losses due to friction or turbulence). This is accomplished by changing the x-component of the velocity, which now is designated by u_2'. Since the actual flow is incompressible, the hypothetical flow cannot possess this property but rather shows distributed sources of which the total strength is Q. Apparently, we have

$$Q = \iint^T (u_2' - u_2)dA,$$

where the integration has to be extended only over the trough since everywhere outside it $u_2' = u_2$. This is designated by the letter T above the integral sign.

Now the momentum theorem in the form (2) will be applied in such a way that first the differences between u_1, v_1, w_1, p_1, and u_2', v_2, w_2, p_2, will be written down and then, as a second step, the differences between u_2', v_2, w_2, p_2, and u_2, v_2, w_2, p_2. The result then is found as the sum of the two partial results given by these steps. For the first step, integral I becomes zero, since $g_2' = g_1$. Integral III causes certain difficulties. It is comparatively small in the case of a pure source flow. Whenever stationary vortices exist, however, as in airfoils, its value may become considerable. Its total value will be denoted by D_i, which will be discussed in detail later. Neglecting contributions of the nature of D_i, the first step applied to integrals I and II leads to a negative drag or driving force $\rho Q u_0$, since there is a distributed source of strength Q between the two planes of integration and there are no losses. Since the integral I is zero, it follows that integral II is

$$\frac{\rho}{2}\int\int (u_1{}^2 - u_2{}'^2)dA = -\rho Q u_0 \doteq -\rho u_0 \int\int^T (u_2{}' - u_2)dA.$$

Applying the second step to integral I, we obtain

$$\int\int^T (g_1 - g_2)dA$$

and, correspondingly for integral II,

$$\frac{\rho}{2}\int\int^T (u_2{}'^2 - u_2{}^2)dA = \frac{\rho}{2}\int\int^T (u_2{}' - u_2)(u_2{}' + u_2)dA.$$

This second step applied to integral III gives a value equal to zero. In total, therefore, the result becomes

$$D = \int\int^T (g_1 - g_2)dA + \frac{\rho}{2}\int\int^T (u_2{}' - u_2)(u_2{}' + u_2 - 2u_0)dA + D_i, \quad (3)$$

where it is seen that the two integrals actually are restricted to the trough. Regarding the third term

$$D_i = \frac{\rho}{2}\int\int [(v_2{}^2 + w_2{}^2) - (v_1{}^2 + w_1{}^2)]dA,$$

the following remark can be made: In case we wanted to treat the y- and z-velocities of a source only, it would be possible to put the front integration plane at an equal distance from the source as the rear plane, which would result in a canceling of the contributions. This relation is not true any more when steady vortices emanate from the body, as in airfoils; in such a case there are velocities in the rear plane of integration to which there are no corresponding ones in the front plane. Then the third term of (3) leads to a resistance D_i different from zero, which is apparently due to the partial vacuum in the vortices. In the three-dimensional wing theory (Art. 114) this resistance is termed "induced drag," and consequently the symbol D_i has been assigned to it. Drag resulting from integrals I and II in the case of airfoils is called "profile drag."

A remark has to be made on non-steady vortices in the turbulent wake. These cause certain deviations, especially in expressions II and III, since the mean values of the products of the velocities do not coincide exactly with the products of the mean values of the velocities.   These deviations are generally small however.   In cases where the wake vortices are more or less regular and strong (Kármán trail, see Art. 82), the previous investigation is not applicable.

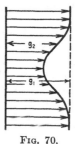

Fig. 70.

The numerical calculation of the resistance in a practical case requires first an experimental determination of g_2 with a Pitot tube.   Outside the trough this value coincides with g_1 which, being the Bernoulli constant of the undisturbed flow, is a constant number (Fig. 70).   Further the static pressure p_2 has to be determined; from this follows $u_2 = \dfrac{\sqrt{2(g_2 - p_2)}}{\rho}$.[1]   For a determination of u_2' it is usually sufficient to interpolate the velocity curves for u_2 by a smooth curve over the trough as indicated by the dotted line in Fig. 70.   It is also possible to calculate it by means of $u_2' = \dfrac{\sqrt{2(g_1 - p_2)}}{\rho}$.

With these measurements, the expressions occurring in integrals I and II are known so that the integrations can be carried out numerically.   The greatest contribution will be furnished by the first integral.   At great distances from the body where the wake velocity is small and the trough is shallow, it leads to the same result as the formula in the Art. 80.[2]   Contribution II gives a correction which becomes rather small at large distances; however, close to the body it is of some importance.   Expression III cannot be calculated in this way since it is not restricted to the trough.   Consequently the method is of use only when the drag consists of profile drag I plus II and when expression III

---

[1] This is true under the assumption that the values of $\dfrac{\rho(v_2^2 - w_2^2)}{2}$ in the trough can be neglected with respect to $g_2 - p_2$, which is admissible in all practical cases.

[2] In case that $u_1 = u_0$, $u_2 = u_0 - u'$, and $p_1 = p_2 = 0$, we have

$$g_1 - g_2 = \frac{\rho}{2}[u_0^2 - (u_0 - u')^2] = \rho\left(u_0 u' - \frac{u'^2}{2}\right),$$

where the second term is small of the second order.   Therefore

$$\iint^T (g_1 - g_2)dA \approx \rho u_0 \iint^T u' dA = \rho u_0 Q.$$

($= D_i$) is without importance.[1] The method has been applied with good results to the measurement of profile drag, especially by Weidinger[2] and Schrenck.[3]

NOTE: The result of the "first step" on page 125 is only equal to $\rho Q u_0$ if the sources of the u_2', v_2, w_2-flow all are located in front of the rear plane of integration. In case, however, that there are sources or sinks behind this plane there will be forces between these and the source Q, which, however, cannot be calculated in such a simple manner. This is probably the reason that Betz's formula becomes inaccurate when applied very close to the body. H. Muttray has shown that Betz's formula (3) can be applied to the flow through a channel of constant cross section A_0. It is only necessary to replace the term $2u_0$ by the sum of the velocity far in front of the obstacle (u_0) and the velocity of the hypothetical flow u_2' far behind the body, i.e., $u_0 + u'_{2\infty}$, where $u'_{2\infty}$ can be set equal to $u_0 + \dfrac{Q}{A_0}$. In this investigation the friction at the channel walls has been neglected.

82. The Kármán Trail.—The phenomena discussed in the last few articles showed a more or less irregular wake behind the obstacle, exchanging its momentum with the neighboring undisturbed water so as to make the wake wider and wider with increasing distance from the body. There are, however, phenomena where the energy is not dissipated directly in an irregular wake but is first transformed into very regular individual vortices. For instance, the two-dimensional flow round a cylinder at certain dimensions and velocities assumes the form shown in Fig. 59, Plate 24. The vortices formed on either side of the body have opposite directions of rotation and form a certain geometrical pattern which is observed quite regularly at some distance behind the obstacle. These vortices do not mix with the outer flow and are dissipated by internal friction only after a long time.

[1] With this procedure the remark made before, about locating the two planes at equal distances from the body and thus eliminating expression III, becomes meaningless. In practical cases no measurements are ever made in the front plane of integration, so that this plane is virtually at infinity where v_1 and w_1 are zero. However the values of v_2 and w_2 are practically always small, so that neglecting them in expression III does not result in any serious error.

[2] WEIDINGER, H., Profile Drag Measurements on a Junkers Airfoil (German), *Jahrb. wiss. Gesellsch. Luftfahrt*, p. 112, Munich, 1926.

[3] SCHRENCK, M., Profile Drag Measurements in Actual Flight by Means of the Momentum Method (German), *Luftfahrtforschung*, vol. 2, No. 1, Munich, 1928.

The phenomenon which had been observed casually by various investigators was first studied experimentally by Bénard.[1] It was reserved, however, to von Kármán[2] to give an explanation of it.

His observations led him to making an investigation of the stability of certain geometrical configurations of these eddies. The calculation, restricted to the two-dimensional case, assumes linear vortices of equal intensity Γ[3] and opposite direction of

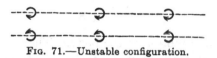

FIG. 71.—Unstable configuration.

rotation lying in parallel rows at equal distances from each other. A further investigation showed that only two different arrangements are possible. The eddies of the one row are situated either exactly opposite those of the other row or they are symmetrically staggered (Figs. 71 and 72). The stability investigation, carried through by means of the method of small oscillations, leads to the result that the first arrangement is unstable with respect to small disturbances while the second pattern is generally unstable also, but becomes stable for a very definite

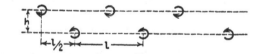

FIG. 72.—Stable when $h/l = 1/\pi$ arc cosh $\sqrt{2} = 0.2806$.

value of h/l (in this case, the pattern is in indifferent equilibrium against disturbances of the wave length $2l$). Von Kármán obtained for the value of h/l the expression

$$\frac{h}{l} = \frac{1}{\pi} \cosh^{-1} \sqrt{2} = 0.2806.$$

Measurements of the distance between the vortices on photographs of actual flows show a good agreement with this calculated

[1] BÉNARD, H., *Compt. rend.* (French), vol. 147, 1908; vol. 156, 1913; vol. 182, 1926; vol. 183, 1926.

[2] VON KÁRMÁN, TH., *Nachr. Ges. Wiss. Göttingen* (German), p. 509, 1911; p. 547, 1912; KÁRMÁN and RUBACH, On the Mechanism of Fluid Resistance (German) *Physik. Z.*, p. 49, 1912.

[3] $\Gamma = \oint \mathbf{W} \cdot d\mathbf{r}$ means the circulation and is a measure for the intensity of the vortex; see p. 207, "Fundamentals," see footnote, p. 3.

value.   The theoretical flow corresponding to the photograph, Fig. 60, Plate 24, is shown in Fig. 73 for a coordinate system which is at rest with respect to the undisturbed fluid.   It is seen that some of the streamlines move in a wavy path between the various vortices, while the other streamlines are closed round the vortex centers.   The total geometrical pattern of the eddies has a velocity of its own,

$$u = \frac{\Gamma}{l\sqrt{8}},$$

in the direction of motion of the body.   The individual vortices gradually remain far behind the body, which moves at a faster rate than the eddies.   The same phenomenon looks entirely differ-

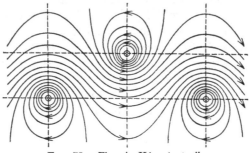

Fɪɢ. 73.—Flow in Kármán trail.

ent to an observer moving with the body, *i.e.*, with respect to a coordinate system in which the body is at rest and in which the velocity of the undisturbed fluid is u_0.   The theoretical stream-line picture for this case can be derived from the one of Fig. 73 by adding to it the constant velocity of the undisturbed flow.

83. Application of the Momentum Theorem to the Kármán Trail.—Von Kármán has shown that by means of the momentum theorem the drag of a body can be calculated from the geometrical pattern of its eddies.   To this end two assumptions have to be made: (1) the actual eddy formation far behind the body should not differ much from the one calculated to be stable (photographic experiments show this assumption to be correct); (2) the fluid at a distance large with respect to the dimensions of the body is assumed to be at rest.   However, it is not possible with Kármán's theory to calculate for any given obstacle the dimensions l and h of the Kármán trail as well as the velocity u

of the eddies. A knowledge of the relation between the dimensions of the eddy pattern and the shape of the obstacle would indeed make the theory of great practical importance. However, it is an advantage to be able at least to calculate the drag when only the dimensions of the vortex trail (by means of a single photograph) as well as the velocity of the eddies are known experimentally. In applying the momentum theorem, it is to be noted that the flow is not steady since there is a periodic eddy formation behind the body. However, the part of the street far behind the body can be considered steady, if the coordinate system is moved with the velocity u of the individual vortices. Choosing a bounding surface enclosing the body and moving forward with this velocity u (Figs. 74 and 75), the fluid enters this surface at the left with a velocity u. At the right the surface

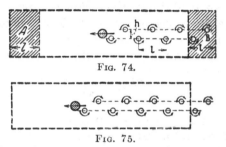

Fig. 74.

Fig. 75.

Figs. 74 and 75.—Application of momentum theorem to Kármán trail.

cuts in between two eddies so that in the Kármán trail there is a velocity to the left (Fig. 73), while outside the flow is to the right with the velocity u.

The body moves with the velocity U relative to the undisturbed fluid, *i.e.*, with the velocity $U - u$ with respect to our system of coordinates. Within the bounding surface new eddies are being formed all the time at the body; moreover, there is the source flow emanating from the body and extending to infinity.

For non-steady motions of this sort the momentum theorem has to be modified because, besides the momentum and pressure integrals over the bounding surface, the change in momentum inside this surface has to be taken into account. According to von Kármán's procedure the phenomenon is considered at the two instants between which just two new vortices have been formed. Since the body moves with the velocity $U - u$ with respect to the trail, and since the distance between two eddies of the same sense of rotation is equal to l, the "period" becomes $T = l/(U - u)$.   Figure 74 shows the initial state and Fig. 75 the final state; the surface of integration is shown in both cases by the heavy-dotted line.   The condition inside the surface of integration at the final state is the same as in the initial state of Fig. 74, if only the shaded area A is omitted from it and the shaded area B is added to it.   If the bounding surface is chosen sufficiently large in the direction of flow as well as across the flow, the contribution of the source flow in the regions A and B is sufficiently small to be negligible.   It is only necessary to calculate the difference in momentum between the undisturbed flow in the region A and the eddy flow in the region B.   According to von Kármán the integral of the x-component of the velocity multiplied by ρ extended over the region B minus the same integral over the region A is equal to $\rho \Gamma h$.[1]   The amount thus calculated is the change in momentum during the time $T = l/(U - u)$. The change of momentum per unit time, constituting a part of the drag, therefore is

$$\rho \frac{\Gamma h}{l}(U - u).$$

The intensity of the source is equal to the jump in the velocity mentioned before, multiplied by h, *i.e.*, $Q = \Gamma h/l$.   According to the investigation of Art. 79 there is a flow of momentum into the long sides of the surface of integration to the amount of $\rho Q u$ which leads to a negative resistance

$$-\rho \frac{\Gamma h}{l} u.$$

The fact that the source moves relatively to the system of coordinates with the velocity $U - u$ does not result in a contribution either in the interior or on the boundary of the surface of integration at least if that surface is sufficiently elongated.

As a third contribution we have the momentum and pressure integrals on the short sides, corresponding to the base surfaces of Art. 79.   One of these bases is in the undisturbed fluid while

---

[1] Since this integration is a linear process, it is permissible to take first the mean value of the velocities, which is found by letting the vortices take all possible positions successively along the length l.   This amounts to an even distribution of the vorticity along l, which causes a uniform jump in the velocity of magnitude Γ/l.   Multiplying this Γ/l by the density and by the area lh leads to the above result.

the other one cuts through the vortex trail. At both those locations the flow is a potential one, and, if the effect of the source traveling with the body is neglected, the flow is, moreover, steady. Therefore the simple equation of Bernoulli can be applied.

It is convenient to calculate at once the sum of the momentum integral and the pressure integral giving the result

$$\rho \frac{\Gamma^2}{2\pi l}.$$

The total drag D per unit length of the cylinder therefore becomes

$$D = \rho \Gamma \frac{h}{l}(U - 2u) + \rho \frac{\Gamma^2}{2\pi l}.$$

Substituting into this expression the values for h/l and for Γ found before, we obtain

$$c = \left[1.587 \frac{u}{U} - 0.628 \left(\frac{u}{U} \right)^2 \right] \frac{l}{d},$$

where d is some linear dimension of the body (for instance, the width of the plate), and

$$D = c \cdot \rho \frac{U^2}{2} \cdot d.$$

The result of the calculation therefore is that the drag is proportional to the square of the velocity, which was to be expected from dimensional reasoning; moreover, the drag coefficient is obtained, which otherwise could have been found only by an experiment. The only limitation is that the drag coefficient is not found directly but rather as a function of the two ratios:

$$\frac{u}{U} = \frac{\text{velocity of the vortex system}}{\text{velocity of the body}}$$

and

$$\frac{l}{d} = \frac{\text{pitch of the vortices}}{\text{characteristic length of the body}}.$$

The ratio l/d can be measured directly from a photograph of the phenomenon, while the ratio u/U can be found from the period of the eddy formation T. A simple analysis shows that this relation is

$$\frac{u}{U} = 1 - \frac{l}{UT},$$

since

$$T(U - u) = l.$$

Von Kármán and Rubach[1] determined the drag coefficient for the cylinder and for the plate from photographs of the eddy trail and from stop-watch measurements of the period. They found $c = 1.60$ for the plate and $c = 0.92$ for the cylinder, with a Reynolds' number between 2,000 and 3,000. Wieselsberger's direct experimental value for the cylinder is 0.93; the most recent measurements of Flachsbart for plates give 1.7; so the agreement in both cases is very satisfactory.

It appears strange that von Kármán's theory, although it assumes an ideal fluid, still takes it for granted that the moving body generates eddies all the time, which is impossible according to classical hydrodynamics. The explanation of this paradox is given by the boundary-layer theory where it was seen that in the limit $\mu = 0$ the fluid can be considered without friction everywhere except in a thin layer adjoining the body. In this layer, which becomes thinner for smaller viscosities, a different limit process has to be performed. In Chap. IV it was seen in detail how this thin boundary layer, where the friction forces cannot be neglected even for fluids with negligible viscosity, is the place where vorticity is created.

84. Bodies of Small Resistance; Streamlining.—Classical hydrodynamics leads to impossible results in all cases where considerable drag is experienced, while for bodies of very small resistance the science can be applied to great advantage. Since in most practical cases, among others those of airplane and airship construction, the problem consists in reducing the drag to a minimum, it appears that a great field is left open for application of the methods of ideal-fluid hydrodynamics. Practical aeronautical construction has derived great help from the modern theories on air motions, *viz.*, the problem of the airship body of least resistance, airfoil theory, and propeller theory.

For an airship body of small resistance, it is essential that the air which is divided at the front closes up again smoothly at the tail end. A shape as indicated in Fig. 49 fits this condition very well. The actual drag, which is very small considering the size

[1] See footnote, p. 131.

of the body, consists practically exclusively of skin friction. According to experiments in the wind tunnel the resistance is about 25 times as small as that of a flat plate of the same size as its greatest cross section.

It is comparatively easy to calculate the flow round bodies of this sort by means of the source-and-sink method of Rankine, as explained in Art. 69 of "Fundamentals."[1] Instead of considering a single concentrated source, as was done in that example, it is more appropriate to take a continuous distribution of sources and sinks along the axis of symmetry. By suitably distributing these sources and sinks, it is possible to obtain a great variety of body shapes, while the thickness of the body is determined by the intensity of the parallel flow superposed on the source-sink flow. It is relatively simple to calculate the body form and the corresponding streamlines for a given source-sink distribution; however, the converse problem, consisting of finding the source-sink distribution for a given symmetrical body, is very much more difficult. This problem will be discussed in the following article.

85. Comparison of the Calculated Pressure Distribution with the Experimental One.—A paper by Fuhrmann[2] deals with the methods of calculating the streamlines round a body for a given source-sink distribution. Besides giving a theoretical calculation, it also reports on measurements of the pressure distribution at the surface of the body. The models experimented with were carefully made so as to resemble the theoretical shapes most accurately. The pressures were measured through very small holes drilled into the hollow models. Figures 76 and 77 show the calculated and measured pressure distributions for two different cases. In general the agreement is very satisfactory; only at the tail end of the body is there an important discrepancy. For the calculated case the pressure at the tail end is equal to the full stagnation pressure, whereas in the actual experiment this cannot be so, since fluid elements cannot penetrate into this high-pressure region on account of their having been retarded in the boundary layer. The integral of the calculated pressures across the whole surface, *i.e.*, the pressure drag, must be zero in all cases, since the calculation has been based on an ideal fluid. The actually observed drag therefore

[1] See footnote, p. 3.

[2] FUHRMANN, G., Theoretical and Experimental Investigations on Balloon Models (German), *Dissertation, Göttingen*, 1912.

exists only in so far as the actual flow deviates from the theoretical one.   For this purpose the drag can be conveniently expressed in terms of an area equal to the side of a cube of the same volume as the airship body under consideration, *i.e.*, $V^{\frac{2}{3}}$, so that

$$D = cV^{\frac{2}{3}}\frac{\rho u^2}{2}.$$

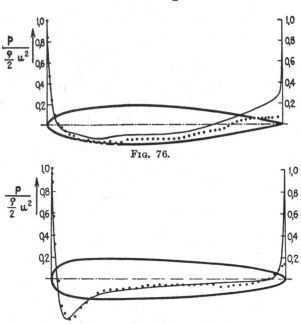

Fig. 76.

Fig. 77.

Figs. 76 and 77.—Pressure distribution on airship hulls.   Full lines are calculated; points measured in wind tunnel.   (*Fuhrmann*.)

With this notation, the calculations based on the experimental pressure distribution gave the following drag coefficients:

Model	I	II	III	IV
c	0.0170	0.0123	0.0131	0.0145

For comparison, the drag coefficients as determined directly in the wind tunnel with the aerodynamic balance are given below:

Model	I	II	III	IV
c	0.0340	0.0220	0.0246	0.0248

It has to be noted that for larger Reynolds' numbers than could be obtained with these experiments, the drag coefficients would have been approximately 30 per cent smaller.   In order to compare these drag coefficients with those of a flat plate of the same surface condition, it is noted that the surface area of the airship models is about equal to seven times $V^{3/4}$, whereas the figures usually quoted for plates refer to the plate area itself. For direct comparison, therefore, the above drag coefficients have to be divided by seven. The experiment shows that for bodies of this streamline form, the total resistance is not materially greater than the friction drag of a flat plate having the same area as the streamlined body. This fact, to a certain extent, can be interpreted as an experimental proof of the theorem of classical hydrodynamics: that the drag of a body in uniform motion is zero.

The converse problem, *i.e.*, the determination of the source-sink distribution for a given shape of body, has been treated by von Kármán[1] on the specific example of the airship ZR-3, which later was named "Los Angeles." The solution was obtained approximately by taking the source and doublet-distribution constant along short sections of the body. He solved not only the case of symmetrical flow but also that of a wind blowing obliquely against the ship. The classical solution for an oblique flow of this sort leads only to a moment tending to place the body perpendicular to the direction of the flow. It definitely does not give a force perpendicular to the flow, *i.e.*, a lift. Since in the actual case lift is obtained, von Kármán assumed in the wake of the airship body the existence of an eddy distribution very similar to that behind an airfoil (Art. 111). This changes the flow markedly, especially around the rear end of the ship, and causes a lift at the front end which is considerably greater than the down push on the tail so that the result is a definite lift, which is in agreement with the experimental facts.

86. Friction Drag of Flat Plates.—When a fluid flows along a flat plate, a force is exerted on the plate in the direction of the flow. It is said that the plate suffers a "friction drag," which as usual is expressed by the relation

$$D = c_f S \frac{\rho u^2}{2},$$

although in this case the drag is proportional neither to the square of the velocity u^2 nor to the area S of the plate. Con-

---

[1] Von Kármán, Th., Calculation of Pressure Distribution on Airship Bodies (German), *Abhandl. Aerodyn. Inst., Tech. Hochschule Aachen*, vol. 6, p. 1, Berlin, 1927.

sequently c_f is no constant but must be a function of the Reynolds' number $R = ul/\nu$ (l being the length of the plate in the direction of flow), as was explained in Art. 59. For small velocities, or rather for Reynolds' numbers smaller than about $5 \cdot 10^5$, the law of Blasius[1] holds that c_f is proportional to the reciprocal of the square root of Reynolds' number

$$c_f = \frac{1.327}{\sqrt{R}},$$

or writing $S = bl$, we have, for $R < 5 \cdot 10^5$,

$$D = c_f bl\frac{\rho u^2}{2} = 1.327\sqrt{\frac{\nu}{ul}} \cdot bl\frac{\rho u^2}{2} = \frac{1.327}{2}\rho bl^{1/2}\nu^{1/2}u^{3/2}.$$

For greater Reynolds' numbers, the experiments by Wieselsberger,[2] Gebers,[3] and Gibbons[4] give a different relation between c_f and R. In the case that $R > 5 \cdot 10^6$ the drag coefficient is proportional to the reciprocal of the fifth root of the Reynolds' number with the proportionality factor 0.072 so that here we have

$$D_f = 0.072\sqrt{\frac{\nu}{ul}} \cdot bl\frac{\rho u^2}{2} = \frac{0.072}{2}\rho bl^{4/5}\nu^{1/5}u^{9/5}.$$

The reason that at a certain Reynolds' number of about $5 \cdot 10^5$ the resistance law suddenly changes is that below this number the flow in the boundary layer along the plate is laminar, while above this number it becomes turbulent, as was explained in Art. 63.[5]

The transition between the two laws is not sharp but very gradual especially for smooth plates with sharp front edges. Figure 78 shows some experimental results obtained by Gebers, Blasius, and Wieselsberger. The last investigator has not used sharp-edged plates but rather rounded ones, where the eddy

[1] BLASIUS, H., Boundary Layers in Fluids of Small Friction (German), *Z. math. Physik*, vol. 56, p. 1, 1908.

[2] WIESELSBERGER, C., Investigations on the Skin Resistance of Canvas Covered Planes (German), *Göttinger Ergebnisse*, vol. 1, p. 120, Munich, 1921.

[3] GEBERS, A Contribution to the Experimental Determination of Drag of Moving Bodies (German), *Schiffbau*, vol. 9, 1908.

[4] GIBBONS, W. A., Skin Friction of Various Surfaces in Air, *1st Ann. Rept., Nat. Adv. Comm. Aeronautics*, 1915, Washington, D. C., 1917.

[5] The limit $5 \cdot 10^5$ is only for very smooth flow; in case the fluid strikes the plate in a somewhat turbulent state this figure is considerably lower.

formation starts immediately at the front end. The shape of the curves shown in Fig. 78 becomes clear when remembering that a sharpened plate where the first part of the boundary layer is still laminar has less resistance than a plate where no such laminar initial boundary layer exists as in the case of Wieselsberger's experiments. This difference between the drags obtained with the two kinds of plates becomes smaller and smaller with increasing Reynolds' numbers on account of the fact that the distance from the front end of the plate where the laminar layer turns turbulent becomes shorter

By making a judicious guess at the ratio between the length of the laminar boundary layer and the length of the plate, Prandtl[1] has found that in the transition region between the

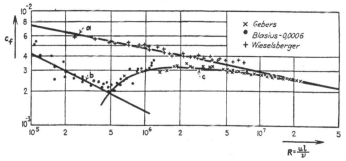

Fig. 78.—Skin friction coefficient *vs.* Reynolds' number for flow along flat plates.

two resistance laws the conditions are well represented by the expression

$$c_f = \frac{0.074}{\sqrt[5]{R}} - \frac{1,700}{R}.$$

In this expression the number 1,700 is still dependent on the degree of turbulence of the fluid coming up to the plate.

It is of interest to mention here the work of Kempf[2] and his associates who have made many investigations on skin friction with special application to ship resistance. A publication of

[1] Prandtl, L., On the Friction Resistance of Air (German), *Göttinger Ergebnisse*, vol. 3, p. 1, Munich, 1927.

[2] Kempf, G., Skin Friction (German), *Werft, Reederei, Hafen*, vol. 5, p. 521, 1925; On the Friction Resistance of Plates of Various Shapes (German), *Proc. Intern. Cong. for Applied Mech.*, Delft, 1924.

Kempf and Kloess[1] investigates the drag of very short plates. The subject of very long plates is treated in the paper by Kempf himself[2] with the appended discussions of Gebers and von Kármán. Other contributions to the subject have been made by Stanton and Marshall,[3] Shigemitsu,[4] and Telfer.[5]

The measurements on long plates show consistently somewhat higher values than those given by the above formulas. In so far as these deviations are due to the influence of very great Reynolds' numbers they form a parallel case to the deviations from Blasius' law for the pressure drop in pipes at high Reynolds' numbers.[6] Based on the experimental results of the flow through pipes, L. Schiller and R. Hermann[7] have made a calculation of the skin friction of plates. For the local drag coefficient c_f' for $ux/\nu > 3 \cdot 10^6$, they give the interpolation formula

$$c_f' = 0.0206\left(\frac{ux}{\nu}\right)^{-0.1294}$$

Integrating this result leads to

$$c_f = \frac{0.024}{R^{0.13}} + \frac{850}{R}.$$

For the case of very smooth flow and a sharp front edge, again an amount $1,700/R$ has to be subtracted from this. As mentioned above, this formula is valid for $R > 3 \cdot 10^6$, while below this Reynolds' number the old formula holds.

Regarding the influence of roughness on the skin friction, it has been found that for surfaces which are not very rough the skin friction is hardly different from smooth surfaces, especially for small Reynolds' numbers. In this case, the roughness inequalities are still within the laminar boundary layer. For larger Reynolds' numbers, where these inequalities protrude

[1] KEMPF, G., and H. KLOESS, Resistance of Short Plates (German), *Werft, Reederei, Hafen*, vol. 6, p. 435, 1925.

[2] KEMPF, G., *ibid.*

[3] STANTON and MARSHALL, On the Effect of Length on the Skin Friction of Flat Surfaces, *Trans. Inst. Naval Arch.*, 1924.

[4] SHIGEMITSU, Skin Friction Resistance and Law of Comparison, *Trans. Inst. Naval Arch.*, 1924.

[5] TELFER, see footnote, p. 104.

[6] See also the note after Art. 48.

[7] SCHILLER, L., and R. HERMANN, Resistance at Large Reynolds' Numbers (German), *Ingenieur Archiv*, vol. 1, p. 391, 1930.

out of the boundary layer on account of its smaller thickness, the rough surface causes considerably greater friction than the smooth one. Surfaces of great roughness, as for instance canvas covered ones, give a practically square resistance law according to experiments of Wieselsberger. This indicates that the drag then is of the nature of a pressure resistance. With increased plate length, c_f decreases in this case also, since with increasing thickness of the boundary layer the relative roughness decreases.

CHAPTER VI

AIRFOIL THEORY

A. EXPERIMENTAL RESULTS

87. Lift and Drag.—The previous chapter dealt with the drag, *i.e.*, with the component of the total force in the direction of the flow. But only in the case of symmetrical bodies where the direction of flow coincides with the axis of symmetry does the direction of the total force coincide with that of the motion. In all other cases there is a definite and sometimes a large angle between them. By decomposing the total force into two components, one in the direction of the flow and another perpendicular to the flow, we are led to the conception of lift. The angle between the direction of the flow and the total force exerted on the body depends very much on the geometrical position of the body with respect to its motion. In practical aeronautics, we are interested in bodies (airfoils) where the total resulting force is nearly perpendicular to the direction of the flow, so that in this case the lift is great and the drag small. The lift serves for carrying the airplane and therefore is a desirable property, whereas the drag is a necessary evil which has to be compensated for by the propeller thrust.

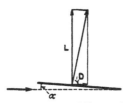

FIG. 79.—Lift and drag on flat plate inclined under 4 deg.; aspect ratio 6.

88. The Ratio of Lift to Drag; Gliding Angle.—It has been known for a long time[1] that a flat plate inclined at a small angle to the direction of the flow has a lift L which is many times greater than its drag D. Figure 79 shows the lift and drag of such a plate inclined under 4 deg. with respect to the direction of flow. The ratio L/D, which is a criterion of the quality of the airfoil, depends not only on the "angle of attack" α but to a great

[1] The oldest literature up to 1902 can be found in Finsterwalder's article on Aerodynamics in the "Encyclopaedie der mathematischen Wissenschaften," vol. IV (Mechanics), No. 17. See also O. Foeppl, Wind Forces on Flat and Cambered Plates (German), *Jahrb. Motorluftschiffstudiengesellschaft*, vol. 4, p. 51, 1910–1911.

extent also on the "aspect ratio." For instance, a rectangle of an aspect ratio 6:1 (*i.e.*, a ratio between the sides of 6:1), shows a considerably larger L/D as a square plate. The ratio D/L is also referred to as the tangent of the "gliding angle," since this is the angle under which the airplane can perform a steady gliding flight.

A far better L/D ratio is obtained by giving the plate a slight curvature or "camber." Figure 80 shows that the lift-drag ratio in this case is about twice as large as for the flat plate.

Still larger lift-drag ratios can be obtained by using regular airfoils as employed on actual airplanes. As an instance, Fig. 81 shows the cross section or "profile" of such an airfoil. It is

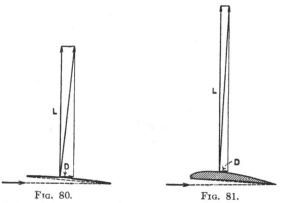

Fɪɢ. 80.   Fɪɢ. 81.

Fɪɢ. 80.—Lift and drag on curved plate inclined under 4 deg.; aspect ratio 6; height of camber equals one twenty-seventh chord.

Fɪɢ. 81.—Lift and drag for airfoil under 4 deg.; aspect ratio 6.

essential that the front end be rounded off nicely and that the top be curved very smoothly; moreover, the tail of the profile should have a sharp edge. The curvature of the bottom side of the profile generally is of less importance. In Art. 91 the relation between the flying characteristics of an airfoil and its profile will be discussed in detail. For good profiles at an aspect ratio of 6:1, it is possible to reach a lift-drag ratio of 20 or more.

Since both the lift and the drag are very much dependent on the angle of attack of the airfoil with respect to the direction of the flow, it is necessary to give a clear definition of this angle. In some cases it is to a certain extent arbitrary which plane through the airfoil is to be taken for the definition of the angle of attack. For profiles where both the top surface and the bottom surface are convex, it is usual to define the angle of attack, as indicated in Fig. 82, by the line connecting the sharp trailing edge with the center of curvature of the nose.

Another method which is common for most airfoils is that of placing a straightedge against the profile as shown in Fig. 83, thus

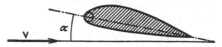

FIG. 82.—Definition of angle of attack for doubly convex airfoils.

defining the angle of attack. This definition has the additional advantage of providing a well-defined point to which the moments of the lift and drag can be conveniently referred. Besides the

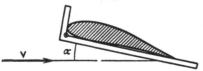

FIG. 83.—Definition of angle of attack and of origin of moments.

usual decomposition of the total air force R into a lift L and a drag D, some other decompositions are used sometimes, for instance, into a tangential force T and a normal force N referred to the center line of the profile.

89. The Lift and Drag Coefficients.—In the preceding chapter the drag coefficient was defined by the formula:

$$C_D = \frac{D}{S\rho\frac{V^2}{2}}.$$

Quite analogous to this, it is possible to define a lift coefficient, namely,

$$C_L = \frac{L}{S\rho\frac{V^2}{2}}.$$

Corresponding coefficients are defined for the other forces sometimes used, for instance: C_N for the normal force, C_T for the tangential force, and C_R for the resultant force. It is necessary to define exactly the area S appearing in the above formulas. In our previous considerations the projected area A in the direction of the flow was used. For airfoils, however, it is common to define S as the largest possible projected area; for instance, for rectangular wings S is equal to the product of the span b and the chord c (Fig. 84); for wings of other shapes we have similarly $S = \int_0^b c\,db.$

90. The Polar and Moment Diagrams of an Airfoil.—Since both the lift and drag depend very much on the angle of attack, it seems logical to plot the lift and drag coefficients as functions of this angle. In the first publications on aeronautics this was usually done, the C_L and C_D values being plotted as functions of

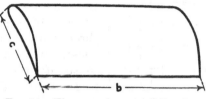

Fig. 84.—The area of an airfoil $S = bc$.

α. It is seen in Fig. 85 that in the region of technical importance from $\alpha = -3$ deg. to $\alpha = 12$ deg., the C_L relation is practically linear and the C_D relation practically quadratic.

In practice, however, the knowledge of the relation between C_L and C_D and the angle of attack α is not necessary, and, moreover, the angle α cannot be easily measured during flight. It

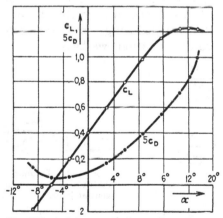

Fig. 85.—Lift and drag coefficients *vs.* angle of attack.

was suggested by Otto Lilienthal to plot C_L as a function of C_D and to write the angle of attack as a parameter into this curve, which is known as the "polar diagram." In Fig. 86 such a diagram is shown with the scale of the drag values five times as large as the scale of the lift values. This is usually done for convenience, since the lift is very much greater than the corresponding drag.

For a full knowledge of the air reaction on the wing, lift and

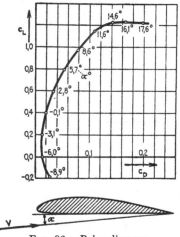

Fig. 86.—Polar diagram.

drag alone are not sufficient since they determine only the magnitude and direction of the total air force but not its location. Instead of specifying a point through which the total air force passes, it is more convenient to specify the moment of this force about a definite point or axis, since the point through which the force acts sometimes lies quite far behind the airfoil. The point O about which the moment is taken usually is the corner point of the straightedge shown in the Figs. 83 and 87.

If N is the normal component of the air force and a its distance from O, the moment is $M = Na$. This moment is considered to be positive when it tends to raise the sharp trailing edge of the wing (Fig. 87). Introducing a moment coefficient C_M with c being the chord of the airfoil, we have

$$C_M = \frac{M}{S\frac{\rho V^2}{2}c} ;$$

and with

$$C_N = \frac{N}{S\frac{\rho V^2}{2}}$$

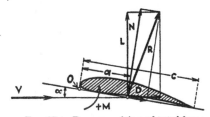

Fig. 87.—Decomposition of total force in lift and drag or in normal and tangential force.

we have

$$a = \frac{M}{N} = c\frac{C_M}{C_N} \quad \text{or} \quad \frac{a}{c} = \frac{C_M}{C_N}$$

It is usual to plot C_M as a function of C_L in the manner shown in Fig. 88, where again the angle of attack is written into the curve as a parameter. Since $C_L = C_N \cos \alpha - C_T \sin \alpha$, it is seen that C_N differs very little from C_L for small angles of attack, so that we can write approximately

$$\frac{a}{c} = \frac{C_M}{C_L}.$$

Therefore if in Fig. 88 the point $\alpha = 3$ deg., for instance, is joined to the origin, and the line thus obtained is intersected with the horizontal line $C_L = 1$, the piece cut off from this horizontal line is approximately equal to a/c. The change of this point of intersection for various angles of attack therefore indicates the travel of the center of pressure along the airfoil.

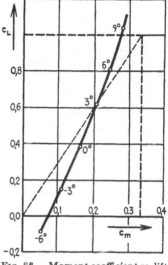

Fig. 88.—Moment coefficient *vs.* lift coefficient.

91. Relation between the Flying Characteristics of Airfoils and Their Profiles.—Since the profiles usually applied in aeronautics cannot be expressed by simple mathematical formulas, a useful and simple classification of them has not yet been devised.[1] Only for a very special class, the so-called Joukowsky profiles (Art. 105), is this possible, since their form can be described completely by two parameters: the thickness and the curvature or camber. Their characteristic properties as a function of these two parameters have been investigated thoroughly.[2]

Geckeler[3] has made an attempt to describe a more general type of profile by means of the theory of complex variables and to find the relations between flying characteristics and profile shape. The first systematic measurements in this direction were

[1] An attempt to express the profiles in mathematical form has been made by E. Everling, An Equation for Airfoil Profiles (German), *Z. Flugtech. Motorluftschiffahrt*, vol. 7, p. 41, 1916.

[2] VON MISES, R., The Theory of the Lift of Airfoils (German), *Z. Flugtech. Motorluftschiffahrt*, vol. 8, p. 157, 1917; SCHRENK, O., Systematic Investigations on Joukowsky Profiles (German), *Z. Flugtech. Motorluftschiffahrt*, vol. 18, p. 225, 1927; LOEW, G., A Contribution to Joukowsky Profiles (German), *Z. Flugtech. Motorluftschiffahrt*, vol. 18, p. 571, 1927.

[3] GECKELER, J., On Lift and Longitudinal Static Stability of Airfoils as a Function of the Profile (German), *Z. Flugtech. Motorluftschiffahrt*, vol. 13, p. 137 and p. 176, 1922.

made in England,[1] where an investigation was made of the influence of a variation in the position of the greatest thickness of the profile on the polar diagram. In the United States systematic experiments have been made on profiles which are stiff enough in themselves to be used without outer struts.[2]

The flying characteristics of a profile are determined by the lift-drag diagram and the moment diagram, which latter gives a measure for the static longitudinal stability. It is to be remembered that the drag of an airfoil depends not only on its profile and on the angle of attack but also to a great extent on the aspect ratio. Therefore in order to determine the influence of the profile shape itself, it is necessary to compare only airfoils of the same aspect ratio. The diagrams given on the following pages are all for an aspect ratio of 5:1.

In general, the total drag of an airfoil can be divided into three parts:

1. The skin friction, which is very much dependent on the condition of the surface of the wing and which can be minimized by making the surface very smooth.

2. A part of the pressure drag which is due to the eddies in the wake behind the wing. This part of the drag is greater for thick profiles than for thin ones.

3. Another part of the pressure resistance which is due to the fact that the air near the wing flows downward on account of the lift and which causes the wing to need a greater angle of attack than would be necessary without this effect.

In Art. 107 it will be explained that this last effect is due to leakage round the wing tips and is the more serious the smaller the aspect ratio is. This part of the drag is equal to the horizontal component of the lift force which is caused by the increased angle of attack. It is called "tip resistance" or "induced drag." In Art. 110 a theory is developed showing that this induced drag is a quadratic function of the lift so that it can be represented by a parabola in the polar diagram, with its apex in the origin but of a shape depending on the aspect ratio. On the diagrams on the following pages, the induced-drag parabola has been plotted always for an aspect ratio of 5:1. The horizontal distance

[1] WIESELSBERGER, C., Investigations on Airfoils in Teddington (German), *Z. Flugtech. Motorluftschiffahrt*, vol. 7, p. 18, 1916.

[2] HERRMANN, H., Aerodynamic Properties of Thick Profiles According to American Tests (German), *Z. Flugtech. Motorluftschiffahrt*, vol. 11, p. 315, 1920.

between this parabola and the polar curve represents the sum
of skin friction and eddy resistance, which sum is known as the
"profile drag."

We shall now discuss the relation between the flying characteristics of airfoils and the shape of their profiles. Figures 89
to 91 show that with approximately equal wing thickness and
equal angle of attack an increase in camber is accompanied by a
considerable increase in lift. However, the drag also increases
even at a faster rate than the lift so that the most favorable
value of L/D becomes somewhat smaller, *i.e.*, more unfavorable
for increasing camber. It is also seen that the moment curve

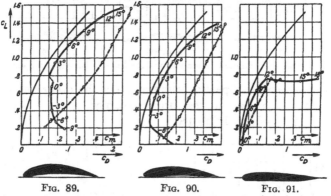

| Fig. 89. | Fig. 90. | Fig. 91. |

Figs. 89–91.—Polar diagrams for airfoils of the same thickness and different
camber.

becomes straighter for decreasing camber, and for a symmetrical
profile it passes through the origin. For such a profile, therefore, the center of pressure does not travel when the angle of
attack is varied.

Figures 92 to 94 show that for airfoils of equal camber the
influence of the thickness is such that the polar curve becomes
flatter and the maximum lift becomes slightly greater for thicker
sections. Figures 95 and 96 show that a thicker profile in general
has a greater drag for the same lift, which is due to the increased
eddy resistance while the skin friction for the two wings is
almost the same. The eddy resistance practically disappears for
very thin profiles and very small angles of attack. Finally, it is
seen in Figs. 97 and 98 that a bending down of the front end of
the profile causes a great increase in the drag for negative angles
of attack.

A rough surface of the airfoil in all cases increases the drag considerably and also diminishes the lift. The most sensitive part of the section in this respect is the front end of the upper

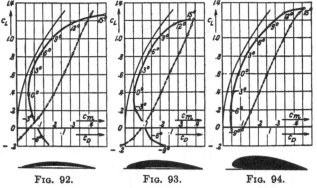

FIG. 92.   FIG. 93.   FIG. 94.

FIG. 92–94.—Polar diagrams for various thickness with the same camber.

side where there is a partial vacuum. On the other hand, considerable roughness on the upper surface near the trailing edge is hardly of any influence.[1]

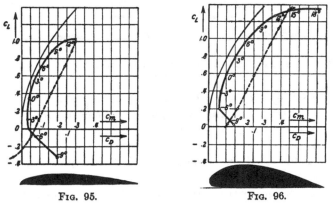

FIG. 95.   FIG. 96.

FIGS. 95 and 96.—The profile drag generally increases with thickness.

92. Properties of Slotted Wings.—Wings with slots have been proposed independently by Lachmann[2] (1918) and Handley-Page[3]

[1] *Göttinger Ergebnisse*, vol. 1, p. 69, 1921; vol. 3, p. 112, 1927.

[2] LACHMANN, G., Slotted Profiles (German), *Z. Flugtech. Motorluftschiffahrt*, vol. 12, p. 164, 1921.

[3] HANSCOM, D., Investigations on Handley-Page Wings (German), *Z. Flugtech. Motorluftschiffahrt*, vol. 11, p. 161, 1921.

(1920) and have acquired considerable importance lately. The main advantage of this type of wing is that the maximum lift obtainable with it is considerably higher than for a normal non-slotted wing. This advantage is to a certain extent offset by the fact that the gliding angle D/L for ordinary horizontal flight is somewhat larger than for simple wings. With slotted wings it is possible to fly horizontally at very much lower speeds which is clearly of great importance for starting and landing. During ordinary horizontal flight it is necessary to close the slot by some mechanical means in order to lessen the drag which otherwise would be prohibitively large. Figure 99 shows a maximum

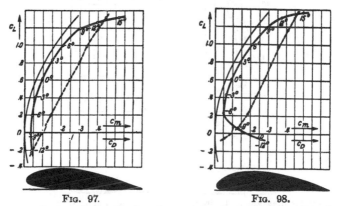

Fig. 97.   Fig. 98.

Figs. 97 and 98.—The profile drag becomes large at negative angles of attack when nose of airfoil is depressed.

lift coefficient for the slotted wing of $C_L = 2.08$, as compared to a value $C_L = 1.38$ for a normal wing. This profile has a maximum lift-drag ratio of 21 as compared to 15 for the slotted wing with closed slot and 13 for the slotted wing with open slot. Still higher lift coefficients (up to $C_L = 2.3$) can be obtained with multi-slotted wings (Fig. 100). For such types the sudden discontinuity in the drag coefficient obtained with single-slotted wings (Fig. 99) does not occur. Naturally, the structural difficulties involved are great.

93. The Principle of Operation of a Slotted Wing.—In order to understand the phenomena in the slotted wing, it is necessary first to comprehend why for a simple wing the lift does not always increase with the angle of attack but starts decreasing when this quantity exceeds a certain limit.

Wing lift is due to a partial vacuum on the top side and an excess pressure on the bottom. This pressure difference must have disappeared at the sharp rear edge where the two currents of air join again. Consequently the pressure along the wing is

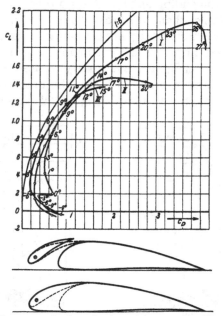

Fig. 99.—Polar diagrams. I, slotted wing with open slot; II, with closed slot; III, normal wing.

increasing toward the rear edge on the top side and decreasing on the bottom side (this will be discussed more fully in Art. 94 and is illustrated by Fig. 102). Bernoulli's equation demands

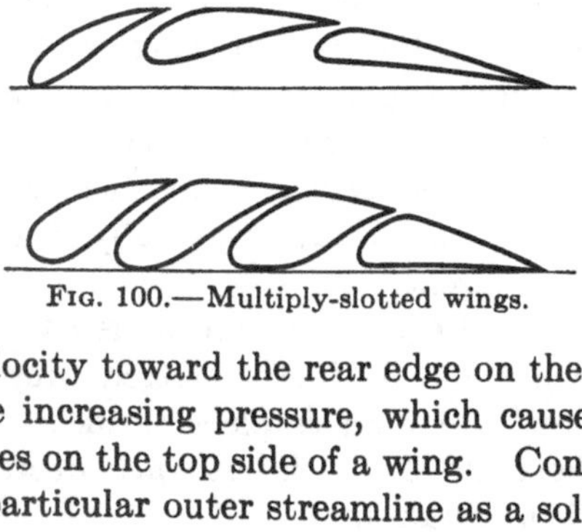

Fig. 100.—Multiply-slotted wings.

a decreasing velocity toward the rear edge on the top side corresponding to the increasing pressure, which causes a divergence of the streamlines on the top side of a wing. Considering for the moment some particular outer streamline as a solid wall and the wing as another solid wall, we have a case of flow through a diverging channel similar to that discussed in the previous chapters. When the angle of attack increases, the lift becomes greater and the pressure gradient on the top side toward the rear edge becomes also greater, which leads to an increase in the angle of divergence of the channel. It was seen before that a conversion of kinetic energy into pressure energy can take place only in diverging channels of a small angle of divergence. As soon as this angle has exceeded a certain rather small value, the flow does not follow the walls any more but breaks away from them and becomes a free jet. This phenomenon occurs also on the top side of a wing for large angles of attack. The flow breaks loose from the wing and the so-called "stalling point" is reached. Because of this effect the streamline picture around the wing is radically changed, and as a consequence of this change the lift decreases. In order to prevent a decrease in the lift it is apparently necessary to prevent the breaking loose of the flow, and this is the function performed by the slot. It was discussed before in Art. 49 that the cause of the breaking loose is a loss of kinetic energy of the particles in the boundary layer due to the action of viscosity. The air coming out of the slot blows into the boundary layer on the top of the wing and imparts fresh momentum to the particles in it, which have been slowed down by the action of viscosity. Owing to this help, the particles are able to reach the sharp rear edge without breaking away. A similar action can be obtained by blowing air at great velocities through little nozzles from the interior of the wing into the boundary layer, as was proposed by Wieland[1] and Seewald.[2]

Another means of preventing the boundary-layer particles from flowing back is to suck them into the interior of the wing in a manner similar to that discussed in Art. 50 for the flow round a cylinder. This is done by means of a blower, and the air thus transported into the wing is blown off at some place where it cannot do any harm.[3]

Still another method of obtaining the same result is to replace the front edge of the airfoil by a rotating cylinder or also by

[1] WIELAND, K., Investigations on a New Kind of Wing with Nozzles (German), *Z. Flugtech. Motorluftschiffahrt*, vol. 18, p. 346, 1927.

[2] SEEWALD, F., Increasing the Lift by Blowing High-pressure Air along the Top Side of an Airfoil (German), *Z. Flugtech. Motorluftschiffahrt*, vol. 18, p. 350, 1927.

[3] SCHRENK, O., see footnote, p. 81.

putting this cylinder inside the wing as shown in Fig. 101. Experiments made by Wolff[1] have shown that airfoils with such a rotating cylinder can be made to have much greater lift coeffi-

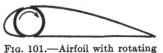

cients ($C_L = 2.43$ with $\alpha = 41.7$ deg.). Of all the methods mentioned, only the slotted wings have been applied so far to practical airplane construction.

FIG. 101.—Airfoil with rotating cylinder.

94. Pressure Distribution on Airfoils.—The pressure distribution on an airfoil is determined experimentally in the same manner as discussed in Art. 85 for airship models. The airfoil model is made of thin metal plate and is hollow inside. At the location where a measurement is to be made, a small hole of approximately $\frac{1}{16}$-in. diameter is provided. The inside of the wing is connected to a manometer by means of a rubber tube. For a test all measuring holes are closed up with putty except the one at which the measurement is to be made. The model is then subjected to the air stream in the wind tunnel and the manometer shows the pressure at the location of the one hole which has been left open. In this manner the entire pressure field on the surface of the airfoil is determined point by point.

Figure 102 shows the distribution in the middle section of an airfoil approximately of the shape of Fig. 86 for various angles of attack. The unit in which all pressures are expressed is the "stagnation pressure." The total force exerted on the wing either on the vacuum or on the pressure side is expressed by the area of the diagram. It is seen in the figure that the major part of the lift is caused by the vacuum action on the top side of the wing. It is also seen in Fig. 102 that the vacuum diagram for an angle of attack of 15 deg. is considerably different from the same diagram of 12 deg. This phenomenon is very intimately connected with the fact that for $\alpha = 14.6$ deg. the lift starts to decrease with the angle of attack (Fig. 86). Such pressure-distribution measurements on the middle section of an airfoil were made for the first time in England in 1911.

Pressure-distribution measurements across an entire wing were also made in England first in 1912–1913,[2] and subsequently

[1] WOLFF, E. B., Preliminary Investigation on the Influence of a Rotating Cylinder in an Airfoil (Dutch), *Verhandel. Rijks Studiedienst Luchtvaart,* Amsterdam, vol. 3, p. 47, 1925; WOLFF, E. B., and C. KONING, Further Investigation, etc. (Dutch), *Verhandel. Rijks Studiedienst Luchtvaart,* Amsterdam, vol. 4, p. 1, 1927.

[2] MUNK, M., *Z. Flugtech. Motorluftschiffahrt,* vol. 7, p. 137, 1916.

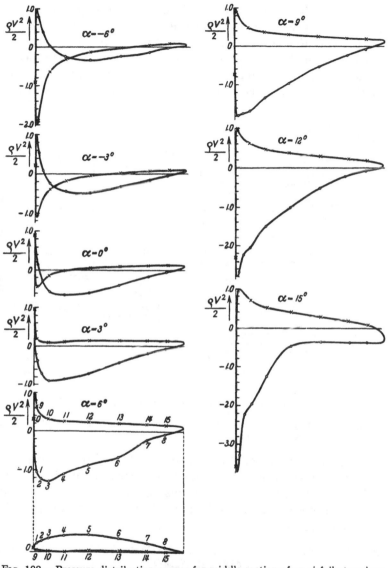

Fig. 102.—Pressure distribution curves for middle section of an airfoil at various angles of attack.

quite often in the United States. From such experiments, the total lift of the wing can be calculated by an integration process first across each section and then along the span (Figs. 103 and 104). The lift and drag values so obtained can be compared to the corresponding values found directly in the wind tunnel by means of the aerodynamic balance. An example of this is shown in Fig. 105, where it is seen that the lift values are in excellent

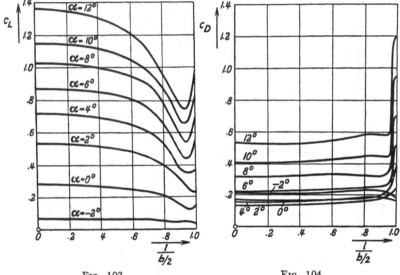

FIG. 103.   FIG. 104.

Fig. 103.—Lift distribution along span (obtained by integrating measured pressures) for various angles of attack.

Fig. 104.—Drag distribution along span for various angles of attack. (Note high drag at tip for large angles of attack.)

agreement, whereas the drag values calculated from the pressure distribution are somewhat lower than those obtained with the balance. The explanation for this naturally is that the skin friction is measured in the balance, but is not included in the calculated figures.

B. THE AIRFOIL OF INFINITE LENGTH (TWO-DIMENSIONAL AIRFOIL THEORY)

95. Relation between Lift and Circulation.—The theory of the lift of a body moving through a fluid is very much different from the theory of drag and offers far less difficulty. The fundamental reason for this is that any explanation of drag requires a consideration of viscosity (even if it is only in a very thin boundary layer), whereas the lift can be explained entirely without the concept of viscosity so that the well-known methods of the classical hydrodynamics of the ideal fluid are applicable. If a body experiences lift, *i.e.*, a force component perpendicular to the flow of the fluid, we can ascribe this phenomenon only to a certain excess pressure on the bottom side of the body and a

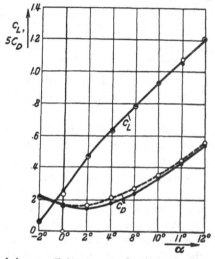

Fig. 105.—Lift and drag coefficients *vs.* angle of attack, from pressure measurements (·) and from aerodynamic balance measurements (○).

certain partial vacuum on the top side (designated by $++$ and $--$, respectively, Fig. 106).

In case this condition is one of steady state, Bernoulli's equation leads to the conclusion that the velocity as an average must be greater above the body than below it. This condition can be explained by superposing on the flow from left to right a circulating flow in a clockwise direction, as was first shown by Rayleigh[1] and Lanchester.[2] This is depicted in Figs. 106 to 108, where it is understood that Fig. 107 shows the purely translatory flow which does not exert any force on the plane but only a turning moment. The superposition of the translatory flow of Fig. 107 and the circulation of Fig. 108 leads to the condition of

[1] Rayleigh, Lord, On the Irregular Flight of a Tennis Ball, *Messenger of Mathematics*, vol. 7, p. 14, 1877, or *Sci. Papers*, Cambridge (England), p. 344, 1899.

[2] Lanchester, F. W., "Aerodynamics," London, 1907.

Fig. 106 where the velocity on top of the plane is greater than on the bottom of it.   The condition of Fig. 107 is the one which exists immediately after starting.   The amount of circulation is designated by Γ which is equal to $\oint \mathbf{w} \circ d\mathbf{r}$ being the line integral of the tangential component of the velocity along any closed

FIG. 106.—Superposition of the
flows of Figs. 107 and 108.

FIG. 107.—Pure translational
flow round inclined plate.

curve surrounding the airfoil.   In Art. 99 it will be explained how the original picture, Fig. 107, is transformed into the one of Fig. 106.

96. The Pressure Integral over the Airfoil Surface.—In the following calculations it is assumed that the airfoil extends to infinity on both sides in the direction perpendicular to the paper.

Owing to this simplification, effects of the wing tips have not to be considered, so that the streamline picture is the same for any cross section perpendicular to the wing, or, in other words, the flow is two dimensional.   The curvature of the airfoil and the angle of attack α with respect to the direction of flow are both so small that it is permissible to assume $\cos \alpha$ as 1.   With this simplification, it is necessary to consider only the x-component of the circulation flow denoted by u_a above the wing and by u_b below the wing.   The directions of u_a and u_b are so indicated in Fig. 109

FIG. 108.—Pure
circulatory flow
round inclined
plate.

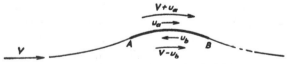

FIG. 109.—Decomposition of velocities.

that both quantities are positive.   The following calculations are made for a span length l cut out from the infinitely long wing.   Denoting the pressures above and below the wing by p_a and p_b, respectively, the following relation is approximately true:

$$L = \iint (p_b - p_a)dS = l \int_A^B (p_b - p_a)dx,$$

where dS is the element of surface equal to $\overline{AB} \cdot l$.   Eliminating from this the pressure by means of Bernoulli's equation:

$$L = \int\int (p_b - p_a)dS = \frac{\rho l}{2}\int^{AB}[(V + u_a)^2 - (V - u_b)^2]dx$$

$$= \frac{\rho l}{2}\int^{AB}[2V(u_a + u_b) + u_a{}^2 - u_b{}^2]dx.$$

With the simplifying assumptions made before, we have

$$\int_A^B u_a dx + \int_A^B u_b dx = \Gamma,$$

which, when substituted into the previous equation, leads to

$$L = l\rho V\Gamma + \int_A^B \frac{l\rho}{2}(u_a{}^2 - u_b{}^2)dx.$$

It will be proved later that the integral appearing in this expression is equal to zero so that the final result is $L = l\rho V\Gamma$, or, in words, the lift is proportional to the circulation.

97. Derivation of the Law of Kutta-Joukowsky by Means of the Flow through a Grid.—In order to derive the above result for the lift in a more exact manner, we shall consider instead of a single wing an infinite number of them forming a "grid." The distance between the individual blades is a, and the coordinate axes are chosen so that x is in the vertical direction positive downward and y is horizontal and positive to the right (Fig. 110). The area over which the integration will be performed consists of a plane $A_1 = al$ on the left side far away from the grid and parallel to the x-axis, a similar plane A_2 to the right of the grid, and two cylindrical surfaces along streamlines.   Since $A_1 = A_2$ the continuity equation leads to

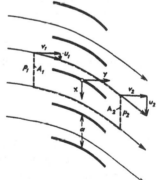

Fig. 110.—Flow through grid for proof of Kutta-Joukowsky lift theorem.

$$v_1 = v_2 = v.$$

Applying the momentum theorem, it is seen that the momentum integral as well as the pressure integral taken over the

two streamlines are neutralizing each other, since the streamlines are identical in shape.   Since, moreover, there is no pressure
component in the x-direction, the momentum theorem applied
to the x-direction becomes

$$-X = \rho l a v(u_2 - u_1).$$

For the y-direction, we find similarly

$$Y = \rho l a(v_1{}^2 - v_2{}^2) + al(p_1 - p_2) = al(p_1 - p_2).$$

Bernoulli's equation, however, gives

$$p_1 - p_2 = \frac{\rho}{2}(v_2{}^2 + u_2{}^2) - \frac{\rho}{2}(v_1{}^2 + u_1{}^2)$$

and, since $v_1 = v_2$, we have

$$Y = \rho l a \frac{u_2{}^2 - u_1{}^2}{2} = \rho l a \frac{u_1 + u_2}{2}(u_2 - u_1).$$

Now the circulation round one blade will be calculated and it
is convenient for this purpose to follow the same contour as was
used for the momentum calculation.   The contributions to
the line integral from the two streamlines neutralize each other
so that we finally obtain

$$\Gamma = a(u_2 - u_1).$$

This substituted in the previous results leads to

$$X = -\rho l \Gamma v,$$
$$Y = \rho l \Gamma \frac{u_1 + u_2}{2}.$$

The individual blades are now moved away from each other until
in the limit they are at an infinite distance apart $(a = \infty)$.
With this process Γ must remain finite and, since $\Gamma = a(u_2 - u_1)$,
it follows that

$$u_2 = u_1$$

or, in words, for a single wing the velocity of the fluid far in
front of it is parallel to the corresponding velocity far behind it.

If the direction of flow at infinity be made to coincide with the
y-axis, we have $u_1 = u_2 = 0$; or, in words, the Y-component of
the force in the direction of the flow vanishes and we obtain
only an X-component, *i.e.*, a lift.   If the x-direction is now

reckoned positive upward and the velocity of the fluid at infinity be denoted by V, we find finally

$$L = \rho l V \Gamma.$$

This very important formula for the lift was first derived by Kutta[1] (1902) and later independently also by Joukowsky[2] (1906).

98. Derivation of the Lift Formula of Kutta-Joukowsky on the Assumption of a Lifting Vortex.—There are several other proofs for this theorem. The proof given by Joukowsky is based on the fact that the flow at great distances from the airfoil is independent of its exact shape. For the stream function he puts in a quite general manner

$$Z = V_z + \frac{\Gamma i}{2\pi} \log z + \frac{A}{z} + \frac{B}{z^2} + \frac{C}{z^3} + \cdots,$$

and, consequently,

$$w = \frac{dZ}{dz} = V + \frac{\Gamma i}{2\pi z} - \frac{A}{z^2} - \frac{2B}{z^3} - \frac{3C}{z^4} - \cdots,$$

where A, B, C, . . . are complex constants. These constants determine the precise shape of the flow round the body and are different for different airfoils and different angles of attack. At large distances (large z) the terms proportional to A, B, C can be neglected, and consequently the velocity field is as if there were only a "lifting vortex" of strength Γ at the origin. It is of importance to note here that this vortex is not a Helmholtz vortex (of which the velocity is zero with respect to the surrounding fluid) but that it is a "lifting vortex" or a "bound vortex" of which the velocity relative to the surrounding fluid is different from zero. It is understood that a lifting vortex is not a physical reality but that it is a very useful concept for the theory of airfoils. The idea of a lifting vortex can be made somewhat plausible by comparing it to a rotating cylinder of which the diameter has shrunk to zero.

With this conception of a lifting vortex the velocity at great distances from the airfoil consists of the superposition of the

[1] KUTTA, W., Lift Forces in Flowing Fluids (German), *Ill. aeronaut. Mitt.*, 1902.

[2] JOUKOWSKY, N., On the Shape of the Lifting Surfaces of Kites (German), *Z. Flugtech. Motorluftschiffahrt*, vol. 1, p. 281, 1910; and vol. 3, p. 81, 1912.

constant velocity V and the secondary circulation velocity
$w = \Gamma/2\pi r$ (Fig. 111).

It is now comparatively simple to derive the lift formula by
means of the momentum theorem, whereby the shape of the integration surface employed determines which part of the total lift
is caused by pressure and which part is caused by momentum.
In case a concentric cylinder round the lifting vortex is used for
integration surface, the pressure integral and the momentum
integral both become $\rho\Gamma V/2 = L/2$, where L is the lift per unit

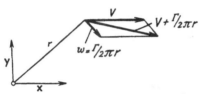

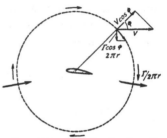

FIG. 111.—The velocity at large distance from airfoil is made up of undisturbed velocity V and secondary velocit
$w = \Gamma/2\pi r$.

FIG. 112.—For circular bounding surface half the lift is due to
momentum and half to pressure.

length of the airfoil. The circulation velocity w due to the
presence of the lifting vortex is $\Gamma/2\pi r$ and everywhere perpendicular to the radius r. From Fig. 112 it is seen that the
momentum integral of the vertical component of the velocity
per unit length of the wing is equal to

$$\rho \oiint dS w \mathbf{w} \cos(\mathbf{n}, \mathbf{w}) = \rho \int_0^{2\pi} r d\varphi V \frac{\Gamma}{2\pi r} \cos^2 \varphi$$
$$= \frac{\rho\Gamma V}{2\pi} \int_0^{2\pi} d\varphi \cos^2 \varphi = \frac{\rho\Gamma V}{2},$$

since

$$\int_0^{2\pi} \cos^2 \varphi d\varphi = \pi.$$

The resulting momentum is directed downward since the velocity
in front of the airfoil has an upward component which is changed
to a downward component behind. The reaction of the fluid,
therefore, gives a lift on the airfoil in an upward direction.

Now we proceed to a calculation of the pressure integral.
Denoting by p_0 the pressure of the undisturbed fluid and by

$V + w$ the geometrical sum of the vectors V and w, Bernoulli's equation is

$$p + \frac{\rho}{2}(V + w)^2 = p_0 + \frac{\rho}{2}V^2.$$

From Fig. 113 it is seen that

$$(V + w)^2 = (V + w \sin \varphi)^2 + (w \cos \varphi)^2.$$

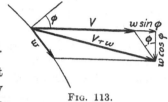

Fig. 113.

Choosing r sufficiently large, it follows that w becomes infinitely small with respect to V, so that the expression w^2 can be neglected. The pressure then becomes:

$$p = p_0 - \rho V w \sin \varphi.$$

Fig. 114.

It is seen immediately that the horizontal component of the pressure integral is zero. The vertical component of p is equal to $p \sin \varphi$ so that the vertical pressure integral per unit length becomes

Fig. 115.

$$\oint p\,dS = \rho \int_0^{2\pi} V w \sin^2 \varphi \, r\,d\varphi =$$
$$\rho r V w \int_0^{2\pi} \sin^2 \varphi \, d\varphi$$
$$= \rho r w \pi V = \frac{\rho \Gamma V}{2}.$$

Pressure below atmospheric pr.

Pressure above atmospheric pr.

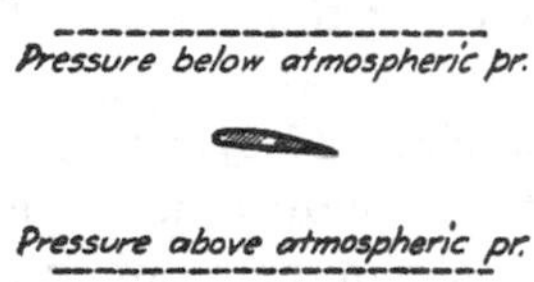

Fig. 116.

Figs. 114–116.—Rectangular bounding surface (114). If the horizontal planes go to infinity, the lift is due entirely to momentum (115). If the vertical planes go to infinity, the lift is due to pressure only (116).

The total lift per unit length is the sum of the pressure integral and the momentum integral and consequently equal to

$$L = \rho \Gamma V.$$

It was said before that the parts taken by the momentum or pressure integrals in the lift depend on the shape of the surface of integration. If, for instance, this surface is taken to be of rectangular form (Fig. 114) and the horizontal sides of the rectangle are moved to infinity, the surface of integration consists of two infinite vertical planes (Fig. 115). In this case the pressure integral is zero and the lift is equal to the momentum integral. On the other hand, if the vertical sides of the rectangle are moved to infinity (Fig. 116), the momentum integral vanishes and the lift is equal to the pressure integral.

In an infinite atmosphere it is therefore entirely undetermined what part of the lift is due to pressure and what part is due to momentum. This is different as soon as the influence of the ground is considered. Then the horizontal sides of the rectangle of Fig. 114 are prevented from moving to infinity, so that only the case of Fig. 116 is possible. This means physically that the lift is always transmitted to the ground in the form of an increased pressure at the surface of the earth. The distribution of this

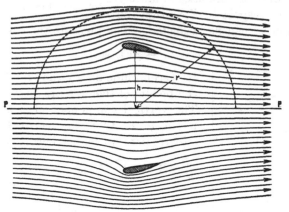

Fig. 117.—Method of images. The lift of an airfoil is transferred to the surface of the earth as increased pressure.

pressure can be conveniently calculated by using the method of mirrored images[1] (Fig. 117).

In that case the earth's surface PP is a plane of symmetry. The surface of integration is taken to be a semicylinder on PP of length l bounded on the bottom by a rectangle cut out of the plane PP. Calculating the pressure integral and the momentum integral across this surface, it is seen that for increasing r, the pressure integral becomes more and more equal to the lift. In the case of an infinite r, the momentum integral converges to zero and the lift becomes equal to the pressure integral.

99. The Generation of Circulation.—It was explained before that a lift can be understood only by assuming a circulation

[1] BETZ, A., Lift and Drag of an Airfoil in the Neighborhood of a Horizontal Plane or of the Earth's Surface (German), *Z. Flugtech. Motorluftschiffahrt*, vol. 3, p. 86, 1912.

flow superposed on the translational flow past the body. How can the existence of such a circulation flow be explained? We assume that at first the fluid is at rest so that the line integral of the velocity along a curve completely surrounding the airfoil is zero, because all velocities are zero. According to the theorem of Thomson (Art. 84, "Fundamentals"[1]), the circulation in a frictionless fluid must remain constant (in this case equal to zero) when the fluid is suddenly put into a uniform translatory motion with respect to the airfoil. This is apparently in contradiction to the experimental fact that there is a circulation round the airfoil.

A close examination of the phenomena shows that the flow in the first moment after starting actually is a potential flow without circulation as shown in Fig. 118 and also in Fig. 48, Plate 19.

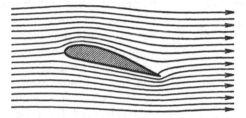

Fig. 118.—Potential flow without circulation.

The most important feature of this potential flow is that the velocity round the sharp rear edge of the airfoil is infinitely large. Owing to the action of the very small viscosity in the boundary layer, however, this large velocity develops into a surface of discontinuity.

This surface of discontinuity emanating from the sharp rear edge is rolled up to a vortex, the so-called "starting vortex." Since this vortex, according to the theorems of Helmholtz, is always associated with the same particles of fluid, it is washed away with the fluid. In the actual experiment with fluids of small viscosity like air or water the flow in the first moment after starting actually shows a great velocity round the sharp rear edge. Immediately afterward, however, a vortex is formed which at once possesses certain finite dimensions, but as in the idealized case this vortex grows rapidly. Figures 49 to 51, Plates 19 and 20, show the generation of such a starting vortex. The same phenomenon is shown in Figs. 52 to 54, Plates 21 and

[1] See footnote, p. 3.

22; in this case, however, the camera is at rest with respect to the undisturbed fluid and the airfoil is moved with respect to it. Owing to the generation of the starting vortex, the velocity field is changed in the sense that a circulating motion is superposed on the translatory motion in such a manner that the circulation round the wing is at any moment equal and opposite to the circulation of the starting vortex. The circulation (round the wing) and consequently the starting vortex increase in intensity until they have reached such a value that the flow joins smoothly at the two sides of the rear edge. As soon as this condition is reached (which usually is the case after the wing has moved about one chord distance), the starting vortex does not increase any more. If the velocity or the angle of attack is increased, another

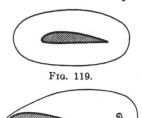

FIG. 119.

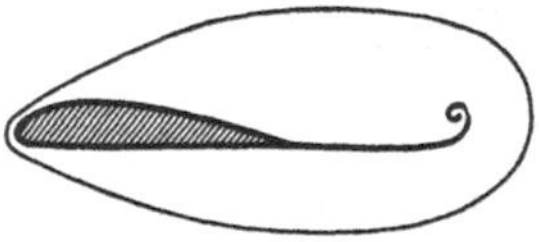

FIG. 120.

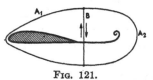

FIG. 121.

FIGS. 119–121.—A fluid line round a wing at rest in various stages of development after starting. The total circulation round it always remains zero, so that the circulation round the airfoil is equal and opposite to that of the starting vortex.

vortex is shed off having the same direction of rotation as the starting vortex. On the other hand, if the velocity or the angle of attack is diminished, a vortex is thrown off which has the opposite direction of rotation as the starting vortex. If the wing is accelerated from rest and immediately afterward stopped, two vortices of equal strength and opposite rotation are thrown off. This phenomenon is illustrated by Fig. 55, Plate 22. Later we shall discuss the energy relations involved in the formation of these eddies.

When it has been understood that the generation of circulation round an airfoil is necessarily accompanied by a starting vortex, it can be shown without difficulty that the existence of circulation is not in contradiction to Thomson's theorem. Trace a line round the airfoil at rest (Fig. 119) and let this line be associated always with the same fluid particles. After the motion has been initiated, this line will always enclose both the wing and the starting vortex (Fig. 120). Since it can be seen in Fig. 55, Plate 22, that the circulation round the airfoil is equal and opposite to that of the starting vortex, it is clear that the line integral round the closed fluid line remains equal to zero. Conversely it can be concluded from Thomson's theorem that the circulation round the airfoil will be equal and opposite to that of the starting vortex. This becomes clear from an inspection of Fig. 121 from which follows that if the line integral round the total closed curve is equal to zero, the line integral round A_1BA_1 must be equal and opposite to the corresponding integral round A_2BA_2.

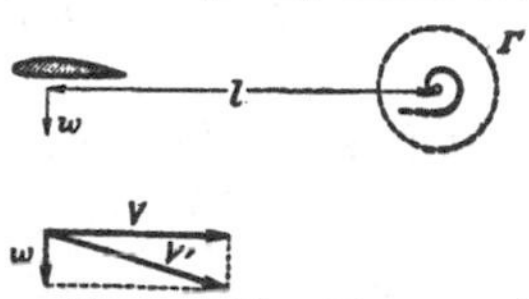

100. The Starting Resistance.—The starting vortex of circulation Γ still influences the flow round the airfoil after it has been washed away from it. From Fig. 122 it is clear that this influence consists of a downward component w of the velocity at the airfoil, having the magnitude

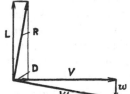

Fig. 122.—Downward velocity w induced by starting vortex.

$$w = \frac{\Gamma}{2\pi l},$$

where l is the distance between the starting eddy and the airfoil. This small additional velocity w causes a deviation in the direction of the relative air velocity and consequently is responsible for a change in direction of the air force. The angle through which this air force is turned is expressed by

$$\varphi = \tan^{-1} \frac{w}{V}.$$

It is clear that this causes a drag, which is large for small distances between the starting eddy and the airfoil. The drag D per unit length of span can be calculated from Fig. 123, as follows:

Fig. 123.—Starting resistance due to downward velocity w.

$$D = L \cos \varphi = L\frac{w}{V} = \frac{\rho V\Gamma}{V} \frac{\Gamma}{2\pi l} = \rho\frac{\Gamma^2}{2\pi l}.$$

It is seen that this starting drag is proportional to the square of the circulation and inversely proportional to the distance of the starting eddy from the airfoil. The work done by this drag when the airfoil moves between l_1 and l_2 consequently is

$$\int_{l_1}^{l_2} D\,dl = \frac{\rho\Gamma^2}{2\pi} \int_{l_1}^{l_2} \frac{dl}{l} = \frac{\rho\Gamma^2}{2\pi} \log \frac{l_2}{l_1}.$$

This work becomes logarithmically infinite when l_2 becomes infinite and it serves to create the kinetic energy round the wing and in the eddy which have equal magnitudes. From Fig. 124 we calculate the kinetic energy K of one of these eddies per unit length of span (perpendicular to the figure), namely,

$$ K = \int_{r_1}^{r_2} 2\pi r dr \frac{w^2}{2} \rho = \frac{\rho \Gamma^2}{4\pi} \int_{r_1}^{r_2} \frac{dr}{r} = \frac{\rho \Gamma^2}{4\pi} \log \frac{r_2}{r_1}. $$

It is seen that the kinetic energy of the circulation is equal

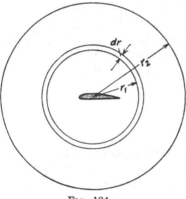

to half the work done on the drag; the other half is accounted for by the kinetic energy in the starting eddy. However, these calculations are pertaining only to the two-dimensional problem of a wing in infinite space. In the real cases, whether of a wing of finite span or of a wing in the neighborhood of the ground, the work done by the drag as well as the kinetic energy of the eddies remains finite.

FIG. 124.

101. The Velocity Field in the Vicinity of the Airfoil.—At first it seems difficult to understand why the direction of the flow far in front is influenced at all by the presence of the airfoil, because the air at such a location has not yet come into bodily contact with the wing. In applying the momentum theorem (page 165) to two parallel vertical planes in front of and behind the wing, it was seen that at a large distance in front of the wing the momentum corresponds to half the lift. This seems even more strange. Lanchester[1] deserves great credit for having given a physical explanation of this phenomenon as early as 1897. He based his considerations on the fact that in order to obtain a lift it is necessary to accelerate air particles downward continuously. If the wing would not move forward, fresh air could be caught and accelerated only by making the wing fall downward (parachute). It is therefore of advantage to make the wing move forward rapidly in order to catch and accelerate new air particles all the time.

[1] LANCHESTER, F. W., see footnote, p. 159.

A simple picture of this downward acceleration of the air under the airfoil can be obtained by letting a flat plate of infinite width fall down freely during a short time. The field of acceleration of the surrounding air can be calculated for this case, and the result of such a calculation is shown in Fig. 125. The air above and below the plate is accelerated downward, while to both sides of the plate the acceleration is directed upward, since the air particles pushed down by the bottom eventually have to get to the top. Now we shall calculate what happens to this field of acceleration when it is moving forward with the velocity V, where V is assumed to be large with respect to the velocities u, v, caused by the acceleration of the plate which is supposed to be of very short duration.

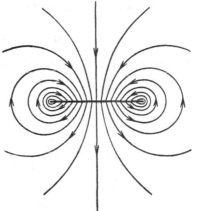

FIG. 125.—Field of acceleration round falling plate (two dimensional).

Further, the displacements of the individual air particles are assumed to be so small that the general expression

$$\frac{Dv}{dt} = \frac{\partial v}{\partial t} + u\frac{\partial v}{\partial x} + v\frac{\partial v}{\partial y}$$

is approximately equal to $\partial v/\partial t$ so that

$$v = \int_{-\infty}^{+\infty} \frac{\partial v}{\partial t} dt.$$

Taking the x-axis of the system of coordinates in the direction of the velocity V and the y-axis in the direction of the acceleration of the plate and introducing the new variable $\xi = x + Vt$, we have

$$\frac{\partial v}{\partial t} = f(x + Vt, y) = f(\xi, y).$$

For a constant x this can be integrated so that

$$v = \int_{-\infty}^{t} f(\xi, y)dt.$$

Moreover for this case of $x =$ constant, we have

$$d\xi = Vdt,$$

so that

$$v = \frac{1}{V}\int_{-\infty}^{\xi} f(\xi, y)d\xi.$$

The velocity of the individual particles in the x-axis therefore is

$$(v)_{y=0} = \frac{1}{V}\int_{-\infty}^{\xi} f(\xi, 0)d\xi.$$

In order to carry out this integration, we use the field of the acceleration in the y-direction for $y = 0$, as shown in Fig. 126.

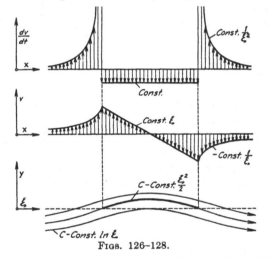

Figs. 126–128.

Fixing the constant of integration so that in the middle of the plate v is zero, the result of Fig. 127 is obtained.

Since the displacements of the particles due to this field of acceleration have been assumed to be small, the velocity v for $y \neq 0$ is approximately the same as for $y = 0$. Now we superpose on the whole phenomenon a uniform velocity $-V$, so that the flow becomes steady. Then, at any instant, $d\xi = dx$. The streamlines for this steady flow can now be calculated from the relation

$$\frac{dy}{dx} = \frac{v}{V + u} \approx \frac{v}{V}$$

or using the above results,

$$y \approx \frac{1}{V}\int v d\xi.$$

It is seen in Fig. 128 that the streamlines above and below the plate have a parabolic shape (since $v \approx \xi$); to either side of the plate, however, the ordinate y decreases with log ξ (because $v \approx 1/\xi$ at a large distance from the plate). From this comparatively simple consideration, it follows that the air in front of the wing is accelerated upward. From the shape of the streamlines it can be concluded that it is of advantage to use a curved plate instead of a flat one. In the case of a parabolic plate the particles in the neighborhood of the plate are accelerated uniformly (see Fig. 126).

102. Application of Conformal Mapping to the Flow Round Flat or Curved Plates.—Independent of Lanchester, Kutta[1]

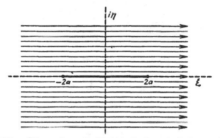

Fig. 129.—Flow along flat plate in $\zeta = \xi + i\eta$ − plane.

calculated the streamlines around an airfoil by means of conformal mapping. The application of conformal mapping to problems of aerodynamics, originated by Kutta in 1902, has proved to be of great value. It should be mentioned, however, that the nature of the method restricts it to two-dimensional problems.

Usually the starting point is the flow around a circular cylinder (see Art. 79, "Fundamentals"[2]). The procedure consists of transforming this cylinder into some airfoil shape by mapping the circle and its streamline picture on another plane by means of a suitable analytic function.

First we consider the trivial case of a uniform flow along an infinitely thin plate. In Fig 129 this plate is represented by a straight line between the points $-2a$ and $+2a$ in the $\zeta = \xi + i\eta$ coordinate system. By means of the function

[1] KUTTA, W., Lift Forces in Flowing Fluids (German), *Ill. Aeronaut. Mitt.*, 1902.

[2] See footnote, p. 3.

$$z = \frac{\zeta}{2} \pm \sqrt{\left(\frac{\zeta}{2}\right)^2 - a^2}, \qquad (1)$$

this straight line is mapped into a circle of radius a and the parallel flow along the straight line maps into the flow round this circle (Fig. 130). In order to verify this we substitute $\zeta = \pm 2a$

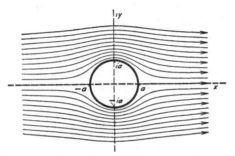

Fig. 130.—Conformal mapping on $z = x + iy$ plane of Fig. 129 by the function:

$$z = \frac{\zeta}{2} + \sqrt{\left(\frac{\zeta}{2}\right)^2 - a^2}.$$

into (1) with the result that $z = \pm a$; also a substitution of $\zeta = 0$ gives for z the values $\pm ia$. In general the points of the ζ-axis between $-2a$ and $+2a$ can be represented by the formula

$$\zeta = 2a \cos \theta$$

which, substituted in (1), leads to

$$z = a \cos \theta \pm \sqrt{a^2 \cos^2 \theta - a^2}$$

or

$$z = a \cos \theta \pm \sqrt{-a^2(1 - \cos^2 \theta)}$$

or

$$z = a \cos \theta \pm ia \sin \theta.$$

Consequently the ζ-axis between $-2a$ and $+2a$ is mapped on a circle of radius a. The two-sheeted Riemann surface of the ζ-plane with the branch cut from $-2a$ and $+2a$ is mapped into the entire $(z = x + iy)$-plane.

By using suitable functions, the flow round the circle in Fig. 130 can be transformed into other flow patterns of great variety. For instance, applying the inverse function of (1)

$$\zeta = z + \frac{a^2}{z}, \qquad (2)$$

it is clear that the original flow of Fig. 129 appears again. With the function

$$\zeta = z - \frac{a^2}{z}$$

the flow round a vertical plane is obtained (Fig. 131). The flow round a plate at an angle α can be derived from (2) by turning the z-plane of Fig. 130 through this angle α (Fig. 132).

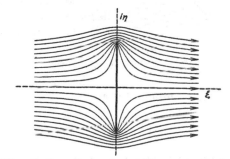

FIG. 131.—Conformal mapping of Fig. 130 by the function:
$$\zeta = z - a^2/z.$$

Kutta also suggested a method for calculating the flow round circular arcs. The dotted circle of Fig. 133 is transformed into the dotted line of Fig. 134 by means of (2). The full circle of

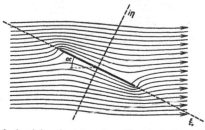

FIG. 132.—Figure derived by first turning Fig. 130 through an angle α, then applying the function of Fig. 131 to it.

Fig. 133 passing through the points $-a$ and $+a$ and having the point if as a center is mapped by the same function into the circular arc of Fig. 134 with a height $2if$. A method for the graphical construction of streamlines round circular arcs based on Kutta's method was worked out by W. Deimler.[1]

---

[1] DEIMLER, W., Constructions of the Kutta Flow (German), *Z. math. Physik*, vol. 60, p. 373, 1912, or *Z. Flugtech. Motorluftschiffahrt*, vol. 3, p. 63, 1912.

103. Superposition of a Parallel Flow and a Circulation Flow.—
In all cases discussed before, the airfoil may experience a turning
moment but it does not have either lift or drag. We already
know that it is necessary to
have a circulation round the
airfoil in order to create lift.
Using the potential function Φ
and the stream function Ψ, as
discussed in Art. 75, "Fundamentals,"[1] it can be shown
with comparative ease that the
flow around a circle of radius a
in the $(z = x + iy)$-plane is

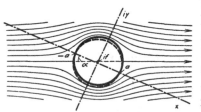

Fig. 133.—Flow round circular cylinder of center $x = 0, y = if$ in a coordinate
system turned through angle α.

determined by the complex function

$$\Phi + i\Psi = V\left(z + \frac{a^2}{z}\right) = Vz + V\frac{a^2}{z}. \tag{3}$$

The first term represents the potential of a parallel flow with the
velocity V, whereas the second term represents a mirroring on
the circle of radius a. The superposition of these two terms
gives the well-known flow round a circular cylinder, as shown in

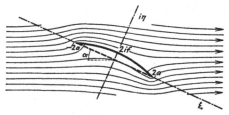

Fig. 134.—Conformal mapping of Fig. 133; the dashed circle becomes the dashed
line and the full circle becomes the arc.

Fig. 135. This can be immediately seen by splitting up (3)
into its real and imaginary components, which is done in the
easiest way by using polar coordinates $z = r \cos \varphi$:

$$\Phi + i\Psi = V \cos \varphi\left(r + \frac{a^2}{r}\right) + iV \sin \varphi\left(r - \frac{a^2}{r}\right).$$

The second term taken equal to a constant is the equation for the
streamlines round the circle. The special case $\Psi = 0$ shows that
the circle itself is a streamline since this equation with $r = a$ is

---

[1] See footnote, p. 3.

satisfied for all values of φ. The same is the case for all values of r different from zero if $\varphi = 0$ or π; in other words, the x-axis is also a streamline.

The stream function of a flow in concentric circles round the circle is given by

$$\Phi_1 + i\Psi_1 = \frac{i\Gamma}{2\pi} \log z,$$

where Γ is constant. Since $i \log z = -\varphi + i \log r$, this is

$$\Phi_1 + i\Psi_1 = -\frac{\Gamma\varphi}{2\pi} + i\frac{\Gamma}{2\pi} \log r.$$

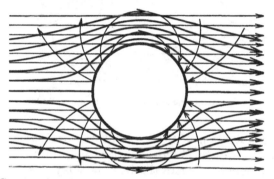

Fig. 135.—Construction of flow round cylinder by superposition of parallel flow and the mirrored image of a parallel flow.

The imaginary term set equal to a constant leads to $r =$ constant; in other words, the streamlines are concentric circles. Differentiation of the function Ψ_1 with respect to r gives for the velocity

$$w = \frac{\Gamma}{2\pi r}$$

or

$$\Gamma = 2\pi r w.$$

It is seen that Γ is the product of the circumference of the circle and the velocity, *i.e.*, the circulation. The numerical value of this circulation will be discussed in the next article, while for the present it is left arbitrary.

By addition of these two stream functions, *i.e.*, by

$$V\left(z + \frac{a^2}{z}\right) + i\frac{\Gamma}{2\pi} \log z,$$

the parallel flow with circulation round a cylinder is obtained. The streamline picture can be easily constructed graphically by drawing the diagonals through the intersections of the individual component flows as is shown in Fig. 136. This figure shows two points of stagnation, the distance of which can be varied by changing the amount of circulation.

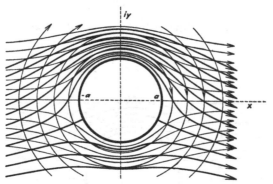

Fig. 136.—Superposition of translational flow round cylinder with circulatory flow.

104. Determination of the Amount of Circulation.—The case of Fig. 136 is of direct practical application to rotating cylinders and also to cylinders where the boundary layer is sucked off on one side only (Art. 52). The greatest value of Fig. 136, however, lies in the fact that it is capable of being mapped into a great number of other profiles by suitable analytic functions. Apply-

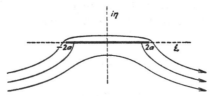

Fig. 137.—Conformal mapping of Fig. 136 such that the circle becomes twice the line from $-2a$ to $+2a$.

ing first the various functions discussed before to the picture of Fig. 136, it is seen in Fig. 137 how the circulatory flow round the cylinder changes its shape when the circle is mapped into a straight line. The same function maps the fully drawn circle of Fig. 138, passing through the points $-a$ and $+a$, and having the point if as center, into the circular arc of Fig. 139. The circulation has been taken such that the two stagnation points just coincide with the points $-a$ and $+a$ of Fig. 138. We now turn our attention to Fig. 133 showing the flow round a circle passing through $-a$ and $+a$ and having the point if for center where the z-plane has been turned through an angle α in the clockwise direction. Superposing on this a flow in concentric circles, it is possible to choose the circulation in such a manner that the point of stagnation at the right just coincides with the

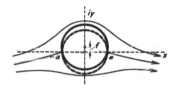

FIG. 138.—Flow round circular cylinder with a circulation such that the two stagnation points are in $-a$ and $+a$.

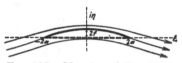

FIG. 139.—Mapping of Fig. 138 by the function $\zeta = z + a^2/z$.

point $+a$. Mapping the z-plane on the ζ-plane by means of (3), the flow round the circle goes into the flow round a circular arc situated obliquely in the flow. On account of a proper choice of the amount of circulation the rear edge of this arc is not encircled, but the flow leaves it smoothly from the bottom as well as from the top side. If this circular arc had been given a greater angle of attack, it would have been necessary to choose a greater circulation in order to secure a smooth flow from the rear edge. This is in agreement with the experimental fact that an increase in the angle of attack causes an increase in the lift and consequently an increase in the circulation.

FIG. 140.—Flow round circular arc with such circulation that rear edge is left smoothly.

The flow sketched in Fig. 140 is quite similar to the one round an actual airfoil with the exception of the large velocity at the sharp front edge.

105. Joukowsky's Method of Conformal Mapping.—In order to circumvent this difficulty at the sharp front edge, Joukowský[1] invented a mapping function by which a circle goes into a profile which is round at the front end. The theoretical consequences

[1] JOUKOWSKY, N., On the Profiles of Airfoils (German), *Z. Flugtech. Motorluftschiffahrt*, vol. 1, p. 281, 1910, and vol. 3, p. 81, 1912.

of this method were discussed later by Kutta.[1]   In Figs. 133 and
134 it was explained how the flow round a circle is mapped into
the flow round an oblique circular arc.   In Fig. 141 the circle
K_1 is surrounded by another circle K_2 in such a manner that the
two circles have only the point $+a$ in common.   The circle K_0
is mapped into the portion $-2a$ to $+2a$ of the horizontal axis,
and the circle K_1 is mapped into the circular arc on this stretch
as a chord.   Consequently the circle K_2 goes into a profile which

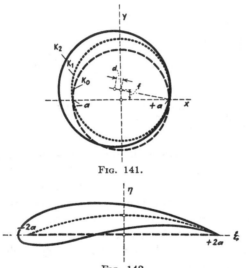

Fig. 141.

Fig. 142.
Figs. 141 and 142.—Mapping of Joukowsky profile.

completely encloses the circular arc and coincides with it only
at the point $+2a$ (Fig. 142).   Since the region outside of the
circle K_2 is mapped on the region outside the new profile, it is
clear that the streamline picture round K_2 is mapped into the
streamline picture round the new profile.   The profiles thus
obtained have great similarity to actual practical airfoils, especially in their front and the middle parts.   However, all Joukowsky profiles show a sharp rear edge without any thickness,
which, of course, in an actual construction cannot be realized on
account of strength considerations.   A simple graphical method

[1] KUTTA, W., On Plane Circulation Flows and Their Applications to Aeronautics (German), *Sitz. Ber. Ak.-Wiss., Math. Phys. Kl.*, vol. 41, p. 65,
München, 1911.

for the construction of Joukowsky profiles was given by Trefftz.[1]

Joukowsky profiles have been investigated theoretically and experimentally a great number of times. The first experiments were carried out by Joukowsky himself in 1912 in his laboratory in Moscow.[2] One year later, Blumenthal[3] calculated the pressure distribution theoretically. He designated the various Joukowsky airfoils by means of three parameters (Figs. 141, 142); first, a, determining the size of the profile; second, f, determining the mean curvature; and third, d, the difference in radii of K_2 and K_1, determining the thickness of the profile. Betz[4] has made a comparison between the calculated values for

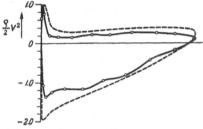

Fig. 143.—Calculated — — and measured ———— pressure distribution round Joukowsky profile.

lift and pressure distribution with the experimental ones. The agreement with the theory is very satisfactory, as shown in Fig. 143, where the full line is the experimental pressure curve and the dotted one is calculated. It is seen that the calculated curve completely encloses the experimental one so that its area, and consequently its lift, is greater than that of the experimental curve. This can be explained by the action of friction. The actual flow does not follow the upper side of the airfoil smoothly but breaks away from it somewhat in front of the rear edge (point A, Fig. 144). The turbulent region thus leaving the tail end of the airfoil causes a loss of lift since the pressure increase at the rear end of the airfoil does not reach the theoretical value.

[1] TREFFTZ, E., Graphical Construction of Joukowsky Profiles (German), *Z. Flugtech. Motorluftschiffahrt*, vol. 4, p. 130, 1913.

[2] JOUKOWSKY, N., "Aerodynamics" (French), p. 145, Paris, 1916.

[3] BLUMENTHAL, O., On the Pressure Distribution along Joukowsky Profiles (German) *Z. Flugtech. Motorluftschiffahrt*, vol. 4, p. 125, 1913.

[4] BETZ, A., Investigation of a Joukowsky Airfoil (German), *Z. Flugtech. Motorluftschiffahrt*, vol. 6, p. 173, 1915.

106. Mapping of Airfoil Profiles with Finite Tail Angle.— Following a suggestion by Kutta[1] (page 77), Kármán and Trefftz[2] have extended Joukowsky's method to profiles of which the tail angle is not equal to zero. In order to accomplish this, it is necessary to use a mapping function which transforms the circle K_0 (Fig. 141) into a figure made up of two circular arcs instead of the mapping function of Eq. (3) which transforms the circle into a piece of straight line.

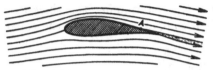

Fig. 144.—Actual flow round Joukowsky profile. Owing to viscosity the flow breaks loose at A.

Kutta's mapping function (1) can be written in the somewhat different form

$$\frac{\zeta + 2a}{\zeta - 2a} = \left(\frac{z + a}{z - a}\right)^2. \tag{4}$$

It was seen that this function maps the circle of radius a of the z-plane into twice the straight stretch from $-2a$ to $+2a$ in the ζ-plane. In particular the singular points $-a$ and $+a$ of the z-plane go into the singular points $-2a$ and $+2a$ of the ζ-plane, and the angle π at the singular points in the z-plane transforms into an angle zero in the ζ-plane.

On the other hand, the function

$$\frac{\zeta + 2a}{\zeta - 2a} = \frac{z + a}{z - a} \tag{5}$$

maps the circle of radius a in the z-plane into another circle of radius $2a$ in the ζ-plane. The angle π in the points $-a$ and $+a$ of the z-plane therefore transforms into angles π in the points $-2a$ and $+2a$ of the ζ-plane.

The exponent 2 in Eq. (4) therefore constitutes the mapping of a circle into a figure consisting of two circular arcs which have degenerated into straight lines, and the angle enclosed by the arcs has also degenerated to zero. Similarly the exponent 1 in

[1] See footnote, p. 180.

[2] Von Kármán, Th., and E. Trefftz, Potential Flow round Airfoil Profiles (German), *Z. Flugtech. Motorluftschiffahrt*, vol. 9, p. 111, 1918.

Eq. (5) transforms a circle into a figure of two circular arcs enclosing angles π. It is to be expected that an exponent k between 1 and 2 would map the circle of the z-plane into a figure consisting of two circular arcs in the ζ-plane enclosing an angle between π and zero. A mathematical investigation shows that this is true. The relation of the angle δ between the circular arcs and the exponent k is given by the relation

$$\frac{2\pi - \delta}{\pi} = k \cdot$$

or

$$\delta = (2 - k)\pi.$$

Therefore, we find that the function

$$\frac{\zeta + 2a}{\zeta - 2a} = \left(\frac{z + a}{z - a}\right)^{\frac{2\pi - \delta}{\pi}} \tag{6}$$

maps the circle of the z-plane into a figure of two circular arcs enclosing the angle δ in the ζ-plane. The function

$$\zeta = z + \frac{a^2}{z},$$

which is identical with Eq. (4), shows that for very great values of z, the points of the ζ-plane approximately coincide with those of the z-plane. Consequently, the velocity at infinity in both planes is equal. On the other hand, the equation

$$2z = \zeta,$$

which is identical with (5), shows that the velocity at infinity in the ζ-plane is twice as great as in the z-plane. In the case of the transformation expressed by Eq. (6), the velocities at infinity in the two planes are also not equal. In order to enforce equal velocities in the two planes, it is necessary to modify Eq. (6) to

$$\frac{\zeta + ka}{\zeta - ka} = \left(\frac{z + a}{z - a}\right)^{k}. \tag{7}$$

An example of the transformation of a Joukowsky circle by means of the function (7) is shown in Figs. 145 and 146. The dotted circle in Fig. 145 is transformed into the dotted figure consisting of two circular arcs shown in Fig. 146. The mean curvature of these two circular arcs would become zero if the center M of the dotted circle K_1 would coincide with the origin of the xy-coordinate system. The greater the distance between M and the origin, the greater the average curvature of the profile. On the other hand, the greater the difference in the radii of the circles K_2 and K_1 the thicker the profile becomes. Besides these two degrees of freedom, the mapping function of Eq. (7) allows a variation in the tail angle. Although a free disposal of three parameters, namely, mean curvature thickness, and tail angle,

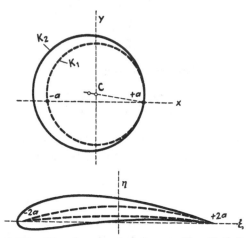

Figs. 145 and 146.—Mapping of a circle to a profile with finite tail angle.

leads to a very great number of profiles, there are certain features in practical airfoils which cannot be approximated by this method. For instance, it fails for profiles having a greater curvature in the front part than in the rear part or having a curvature of which the center line shows an inflection point (S-profiles). Kármán and Trefftz therefore discuss an approximate method by which it is possible to map the circle on any airfoil section. By applying modern theorems of the theory of functions of complex variables, von Mises[1] and W. Müller[2] have made great progress in this direction as well as in the calculation of the relation between the lift and the angle of attack.

[1] Von Mises, R., The Theory of Lift of Airfoils (German), *Z. Flugtech. Motorluftschiffahrt*, vol. 8, p. 157, 1917; vol. 11, pp. 68 and 87, 1920; *Z. angew. Math. Mech.*, vol. 2, p. 71, 1922.

[2] Müller, W., On the Theory of Mises' Profile Axes (German), *Z. angew. Math. Mech.*, vol. 4, p. 186, 1924; Müller, W., "Mathematical Hydrodynamics" (German), Berlin, 1928.

C. THREE-DIMENSIONAL AIRFOIL THEORY[1]

107. Continuation of the Circulation of the Airfoil in the Wing-tip Vortices.—With a wing of infinite length the various pressures above and below it are constant along the span because the pressure distribution is the same for all vertical planes. This condition cannot exist for wings of finite width since here the pressure differences between the top and bottom have to disappear gradually toward the wing tips. On account of the greater pressure below the wing surface than above it, some air will flow from the bottom to the top

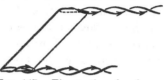

FIG. 147.—Tip vortices leaving a finite wing.

round the wing tips. Therefore a sidewise current exists over most of the wing surface, directed outward on the bottom side and toward the center on the top side. This causes a surface of discontinuity in the air leaving the wing, which is ultimately

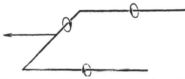

FIG. 148.—Simple picture of vortex system of finite wing.

rolled up into two distinct vortices. According to the theorem of Helmholtz, these vortices always consist of the same air particles so that they leave the wing approximately with the velocity V in the form of two lines as shown in Fig. 147. In order to simplify the calculations, it has been assumed that the circulation, and consequently the lift, remains constant along the entire length of the wing and diminishes suddenly to zero at the tips. It was seen before that the flow round an infinitely long airfoil can be replaced by a flow due to a linear vortex in the wing. This is permissible also for a finite wing so that the simplest picture of the situation is given by three linear vortices as shown in Fig. 148. This phenomenon can be explained in a somewhat different manner: a linear vortex cannot terminate in the interior of the fluid but only at infinity or at a surface. It is clear therefore that the "bound" vortex of the wing cannot end at the tips but must be continued to infinity as a "free" vortex. If the airplane has started some place, the starting vortex closes

[1] PRANDTL, L., Airfoil Theory (German), Pts. I and II, *Nachr. Ges. Wiss.*, *Göttingen*, p. 151, 1918, and p. 107, 1919.

the two long free-vortex lines at the other end so that the total
vortex picture consists of a very long rectangle.

**108. Transfer of the Airplane Weight to the Surface of the
Earth.**—Although the condition of Fig. 148 is only a very rough
approximation of the actual flow, it is capable of explaining
some of the consequences at large distances from the airfoil.

One of the most useful results that can be drawn from it is the
law according to which the weight of the airplane is transferred
to the surface of the earth.   The introduction of a mirrored
image, as used before on page 166, will prove useful also in this
case, since then the normal velocity component at the earth

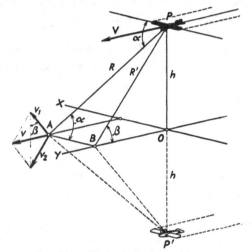

Fig. 149.—Method of mirrored images.

automatically reduces to zero.   In order to obtain a steady flow,
the coordinate system is chosen such that the airplane appears
to be at rest, the x-axis being in the direction of the wing, the
y-axis in the direction of motion, and the z-axis pointing downward (Fig. 149).

The velocities u, v, w of the vortex system are supposed to be
so small compared with the velocity V that their squares can be
neglected with respect to the products of any of them with V.
Denoting by p_0 the pressure of the undisturbed air and by p the
difference between the actual pressure and p_0, Bernoulli's equation gives

$$p_0 + p + \frac{\rho}{2}\{u^2 + (v - V)^2 + w^2\} = p_0 + \frac{\rho}{2}V^2,$$

or neglecting the quadratic terms in u, v, and w,

$$p = \rho V v.$$

The pressure p at the earth's surface (*i.e.*, the increased pressure due to the presence of the airplane) can be calculated by first determining v as a function of x and y for $z = 0$, then integrating v over the xy-plane and multiplying the result by ρV.

The two "free" linear vortices leaving the wing tips are sloping downward slightly on account of their own field of motion. This slope is so small that it will be neglected in the present calculation. Then the wing tip vortices do not contribute anything to v, which is caused entirely by the bound vortex. The length of this bound vortex b is equal to the span of the wing, but for an actual calculation a somewhat smaller value than this is more appropriate in order to take account of the fact that a part of the free vortices leaves the wing between the tips. Denoting the circulation round the airfoil by Γ and observing that b is small with respect to the height h of the plane above the ground, we find

$$v_1 = \frac{\Gamma b \sin \alpha}{4\pi R^2},$$

where the direction of v_1 is perpendicular to the plane ABF, Fig. 149.

The mirrored image of the airplane leads to a corresponding velocity v_2 so that the actual velocity v on the earth becomes the geometrical sum of v_1 and v_2. With the angles α and β as defined by Fig. 149, we have

$$v = 2v_1 \sin \beta = 2v_1 \frac{h}{R'}$$

or, since

$$\sin \alpha = \frac{R'}{R},$$

this becomes

$$v = \frac{\Gamma b h}{2\pi R^3}.$$

Eliminating the circulation by means of the relation $L = \rho \Gamma V b$, we find that

$$p = \frac{Lh}{2\pi R^3}.$$

The increased pressure therefore is seen to have rotational symmetry with respect to the location of the airplane. Right under the plane, the maximum pressure

$$p_{\text{max.}} = \frac{L}{2\pi h^2}$$

exists; it is seen that even for small heights, this pressure is extremely small since the height appears squared in the denomi-

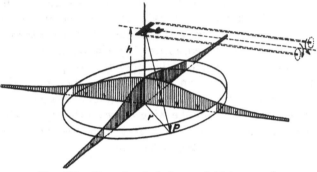

FIG. 150.—Transfer of airplane weight to ground.

nator. The pressure distribution is shown schematically in Fig. 150. Calculating the pressure integral over the entire surface of the earth, using the notations of Fig. 151, we get

$$\iint_{-\infty}^{\infty} p\,dx\,dy = \int_{r=0}^{\infty}\int_{\epsilon=0}^{2\pi} p\,dr \cdot r\,d\epsilon.$$

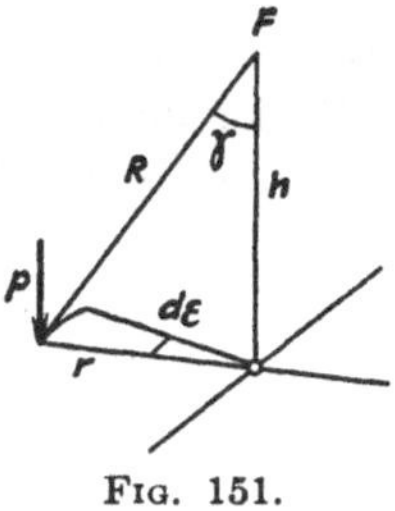

Fig. 151.

Noting that $\cos \gamma = h/R$ and $r = h \tan \gamma$ so that

$$dr = \frac{h\,d\gamma}{\cos^2 \gamma}$$

the integral becomes

$$\frac{L}{2\pi}\int_{\gamma=0}^{\frac{\pi}{2}}\int_{\epsilon=0}^{2\pi} \frac{h^3}{R^3}\frac{\sin \gamma}{\cos^3 \gamma} d\gamma\,d\epsilon = L\int_0^{\frac{\pi}{2}}\sin \gamma\,d\gamma = L.$$

It is seen therefore that the lift L is completely carried by the ground in the form of increased pressure.

109. Relation between Drag and Aspect Ratio.—The bound vortex finds its continuation in two tip vortices extending along the entire flight path of the airplane. These tip vortices contain a certain amount of kinetic energy that has been created by the plane in its flight, which presupposes that the plane must have been doing work. This means that the plane moving through an ideal fluid has to overcome a certain drag. In this connection we do not consider the profile drag (Art. 91) which, in addition, always exists in an actual fluid.

Other things being equal, the lift is proportional to the wing span; on the other hand, the kinetic energy of the tip vortices, and consequently the drag caused by them, is approximately independent of the span. It is seen therefore that this drag is much more important for short spans than for long ones, or the

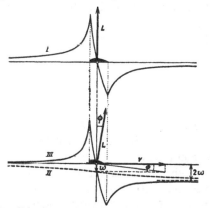

Figs. 152 and 153.—Distribution of vertical velocity. I, for infinite wing; II induced by tip vortices of finite wing; III = I + II total for finite wing.

drag per unit of lift is greater for a wing of small aspect ratio than for one of large aspect ratio. Since it is very important for gliders to have a favorable ratio of lift over drag, wings of large aspect ratio are used. A theoretical calculation of this part of the drag is possible by assuming the velocities due to the various vortices to be small with respect to the velocity V of the wing, which is practically always the case.

110. Rough Estimate of the Drag.—First an estimate of the drag will be obtained by means of momentum and energy considerations without entering into the details of the flow. The wing-tip vortices cause a downward motion of the air at the wing, which will be shown to be responsible for the drag.

Figure 152 shows the distribution of the vertical component of the velocity along a line in the direction of flight for a wing of infinite length. A similar picture was calculated theoretically for an idealized case in Figs. 126 to 128. In Fig. 152 we see, what we knew before, that the air in front of the wing is deflected upward and is then pushed downward by the wing. Now suppose a finite piece to be cut from the infinite wing of Fig. 152. The tip vortices will cause an additional downward component w of the air, of which the distribution is indicated by the dashed line II, Fig. 153. This, however, so influences the direction of the wind velocity at the wing, that it appears to be flying in a direction inclined with the angle $\varphi = \tan^{-1} \dfrac{w}{V}$ with respect to the actual direction of flight. Since the air force is perpendicular to the apparent direction of motion, it includes the angle φ with the actual direction of motion. If w_0 is the downward velocity at the center of pressure of the wing due to the action of the tip vortices, we have the relation

$$\frac{D}{L} = \frac{w_0}{V}. \tag{1}$$

The induced velocity w_0 is not the same for all points along the span; outside the wing tips the velocities are even directed upwards. This complication is neglected at first and the simplifying assumption is made that a certain amount of fluid passing through a certain cross section S' is influenced by the wing in such a manner that it acquires a downward velocity w_1, while the rest of the air does not experience any downward deviation at all. The area as well as the shape of this cross section S' cannot be determined by this rather rough reasoning, and its calculation must be postponed to the more refined analysis which is to follow later. Anticipating these later calculations, it is mentioned here that S' does not depend on the angle of attack.

According to our assumptions, a mass of air $\rho S' V$ is given the downward velocity w_1, which causes as a reaction the lift

$$L = \rho S' V w_1.$$

An application of the energy theorem stating that the work done on the drag is equal to the kinetic energy created in the unit of time leads to

$$DV = \rho S' V \frac{w_1{}^2}{2}.$$

Substituting the expression for the lift obtained above, we get

$$D = \frac{Lw_1}{2V} = \frac{\rho S' V}{2V}\left(\frac{L}{S'\rho V}\right)^2 = \frac{L^2}{4\frac{\rho}{2}V^2 S'}.$$

Comparing this result with Eq. (1), it is seen that

$$w_0 = \frac{w_1}{2};$$

or, in words, the downward velocity due to the tip vortices at the center of pressure of the wing is half as great as the final downward velocity far behind the wing (see also Art. 113).

The calculation of the drag is now reduced to a determination of the area S'.

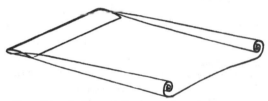

Fig. 154.—Surface of discontinuity behind a wing.

111. The Jump in Potential behind the Wing.—In order to penetrate further into the problem, it is necessary to drop the simplified picture of Fig. 148 and to investigate the wake of the wing in detail. With the two-dimensional flow round an infinite wing there is a "surface of separation" behind the trailing edge, separating the air which has passed over the wing from that which has passed under it. The velocity on both sides of this surface however is the same, so that it has hardly any physical reality. With the finite wing, however, there is a lateral flow toward the tips below the wing and toward the center above it. This flow continues in the surface of separation so that it becomes a surface of discontinuity for the velocity (see page 221, "Fundamentals"[1]). Owing to the action of the tip vortices this surface moves downward with a velocity w_1, which increases with the lift. A detailed consideration shows that this surface rolls itself up in the manner shown in Fig. 154. However, the rate of rolling up is small for a small w_1 and in the following deliberations it will be neglected, presuming the surface of discontinuity to be a plane. In other

[1] See footnote, p. 3.

words, the flow is the same in all planes perpendicular to the direction of flight; the phenomenon is a function of x and z only, independent of y (Fig. 155). This is a two-dimensional problem, which can be solved with the methods of classical hydrodynamics.

We now return to Lanchester's conceptions on page 171 and Fig. 126. Instead of a downward acceleration of the wing itself in its various consecutive positions we shall consider an instantaneous acceleration (impulse) of the whole surface of separation. In other words, this surface of separation is momentarily solidified into a "board," and this board is given a downward impulse. (In order to take care of the variation of w_1 with x, the "board" may be considered elastic.)

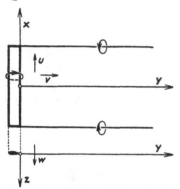

Fig. 155.—Surface of discontinuity in the case of small induced downward velocity w.

The two-dimensional field of acceleration thus obtained has the distribution of Fig. 125. During the short interval τ of the acceleration, let $-p_a$ be the decrease in pressure on the top of the board and p_b be the increase on the bottom side. The "impulse pressure" at each point of the air is then given by

$$\int_0^\tau p\,d\tau,$$

where p is the pressure difference with the undisturbed state. The general equation of Bernoulli for non-steady motions applied to a coordinate system at rest with respect to the undisturbed air is

$$\frac{\partial \Phi}{\partial t} + \frac{1}{2}w^2 + \frac{p}{\rho} = \text{const.}$$

With our assumption that w is very small, this simplifies to

$$\frac{\partial \Phi}{\partial t} + \frac{p}{\rho} = \text{const.}$$

In order to determine the constant, the phenomena at a great distance to the side of the "board" will now be considered. Here the fluid is hardly disturbed and both terms on the left-hand side go to zero in the limit when the distance to the side increases indefinitely so that the constant also becomes zero. Therefore

$$\frac{\partial \Phi}{\partial t} = -\frac{p}{\rho}.$$

On account of the very short duration of the impulse of the "board" the "substantial" integration can be replaced by the "local" one. Assuming incompressibility of the air, the previous expression can be integrated:

$$\Phi_r - \Phi_0 = -\frac{1}{\rho}\int_0^\tau p\,dt.$$

Since the motion started from rest and consequently $\Phi = 0$ at the time $t = 0$, we have

$$\int_0^\tau p\,dt = -\rho\Phi_r,$$

where Φ_r is the potential of the flow after the impulse, which remains constant from then on.

In order to calculate the total impulse for a flight path of length l (perpendicular to the plane of the drawing of Fig. 125), the impulse pressure has to be multiplied with the area so that it becomes

$$\int_0^\tau dt \int_{x_1}^{x_2} l(p_b - p_a)\,dx = l\rho \int_{x_1}^{x_2}(\Phi_a - \Phi_b)\,dx.$$

The integral

$$\int_{x_1}^{x_2} l(p_b - p_a)\,dx$$

is the expression for the force on the area $(x_2 - x_1)l$ at any instant between zero and τ. After the impact, *i.e.*, for $t > \tau$ we have $p_b = 0$ and $p_a = 0$, and the integration from zero to τ gives the total impulse. According to the principle of action and reaction, this impulse acting on a surface of the length l is equal to the lift on the airfoil multiplied by the time T, which the wing needs in order to fly the distance l $(T = l/V)$. Consequently,

$$l\rho \int_{x_1}^{x_2}(\Phi_a - \Phi_b)\,dx = LT = L\frac{l}{V}$$

or

$$L = \rho V \int_{x_1}^{x_2}(\Phi_a - \Phi_b)\,dx.$$

It will now be explained that the potential jump at the surface of separation is tied up with the circulation round the wing. The line integral of the velocity along any curve which does not pierce the surface of separation behind the wing must be equal to zero since the curve lies entirely in a region of potential flow.   If the points B and A of the curve in Fig. 156 are made to converge to a common point in the surface of separation, the single curve falls apart into two branches.   Since the line integral along the branch enclosing the wing must be equal to its circulation Γ, the corresponding integral along the other branch of the curve must be equal to $-\Gamma$ and must also be equal to $\Phi_a - \Phi_b$.   Substituting this result into the last formula, we get

$$L = \rho V \int_{x_1}^{x_2} \Gamma dx. \tag{2}$$

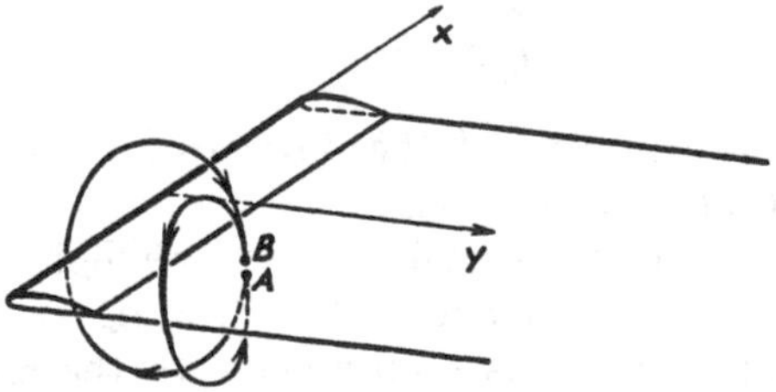

FIG. 156.—The line integral of the tangential velocity round the closed curve is zero as long as it does not pierce the surface of discontinuity.   When the curve falls in two branches by letting $B = A$, the circulations of the two branches are equal and opposite.

Assuming the circulation and consequently the lift to be constant along the x-axis, we obtain

$$L = \rho V \Gamma (x_2 - x_1) = \rho V \Gamma b,$$

where $b = x_2 - x_1$ denotes the span of the wing.   This constitutes another proof of the lift theorem of Kutta-Joukowsky, and it is seen from Eq. (2) that this theorem is still applicable for a circulation which varies along the span.

In order to proceed with a calculation of the area S', discussed on page 190, which is essential for a determination of the drag, we replace the potential Φ by the expression $\Phi/w_1 = \varphi$, which has the dimension of a length.   With this notation, the integral $\int_{x_1}^{x_2} (\varphi_a - \varphi_b) dx$ is of the dimension of a surface, and φ obviously can be regarded as the potential of a similar flow in which the velocity $w_1 = 1$.   Comparing the expression for the lift,

$$L = \rho V w_1 \int_{x_1}^{x_2} (\varphi_a - \varphi_b) dx,$$

with the expression derived from the momentum theorem,

$$L = \rho V w_1 S',$$

the area S' is seen to be equal to

$$S' = \int_{x_1}^{x_2} (\varphi_a - \varphi_b)dx.$$

The integration can be performed as follows:

$$\int_{x_1}^{x_2} (\varphi_a - \varphi_b)dx = \int_{x_1}^{x_2} \varphi_a \, dx + \int_{x_2}^{x_1} \varphi_b dx = \oint^{x_1x_2x_1} \varphi dx,$$

whereby the path of the last integral extends round the surface of separation taking first the φ-values on the top side and then on the bottom side. With this understanding, we write

$$L = \rho V w_1 \oint \varphi dx$$

and

$$S' = \oint \varphi dx.$$

112. The Vortex Sheet behind the Wing with Lift Tapering Off toward the Tips.—It has been assumed until now that the lift is constant along the span, and that the circulation which drops suddenly from its value Γ to zero at the wing tips is continued in two free tip vortices, moving away from the wing with the air.

If we consider the more real case of a lift distribution which has a maximum at the middle of the span and which decreases gradually to zero toward the tips, it follows that free vortex lines emanate from the rear edge of the wing in all places where the circulation changes along the span. Let the circulation and consequently the potential difference be Γ at some point A_1 of the airfoil (Fig. 157) and let the circulation at A_2 be

$$\Gamma + \frac{\partial \Gamma}{\partial x}dx.$$

The line integral of the velocity along the closed curve of Fig. 157 is zero since the curve remains entirely inside the region of potential flow. Letting the points $1'$, $1''$, $2'$, $2''$ converge toward the surface of separation, the value for the circulation of the small closed loop of Fig. 158 is

$$\frac{\partial \Gamma}{\partial x}dx.$$

This is the circulation of the vortex leaving the airfoil at that particular location. If the circulation is variable along the entire span and is not constant at any point, the vortex lines leaving the wing form a surface which is identical with the separation surface discussed before. It is sometimes referred to as a "vortex band."

This result can still be interpreted in the following manner. The lift being a maximum in the middle of the span decreases gradually toward the tips. Since the lift is made up of a partial vacuum above the wing and an excessive pressure at the bottom side of it, the decrease in lift toward the wing tips is associated

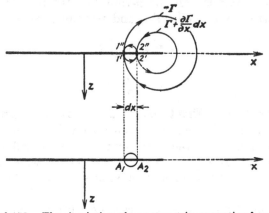

Figs. 157 and 158.—The circulation of a vortex strip emanating from the wing is
$$\frac{\partial \Gamma}{\partial x}dx.$$

with an increasing pressure on the top and a decreasing pressure at the bottom toward the tips. Under the influence of these pressure differences along the span, the air particles flowing by from front to rear are pushed to the side somewhat, namely, toward the middle on top of the wing and toward the tips at the bottom. With a non-viscous fluid, the air stream, which is divided into a top and bottom stream at the front end of the wing, is closed up again at the rear end. Due to the lateral components of the velocities, there will be a discontinuity in the velocity at the separation surface where the two streams come together. The differences in lateral velocity at this separation surface increase with increasing lateral pressure variation and with increasing change in the circulation Γ. The absolute values of the velocities above and below the surface of separation must be equal according to

Bernoulli's equation, since the pressures on both sides of it are equal. Figure 159 shows schematically the distribution of the vorticity in the surface of separation if the circulation increases stepwise from the wing tips toward the center.

In addition to the velocity field of the bound vortex we have to consider the velocities caused by the free vortices leaving the

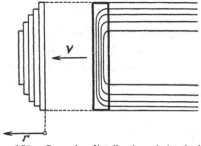

Fig. 159.—Stepwise distribution of circulation.

trailing edge of the wing. It was seen before that the induced drag is due to the downward velocity at the airfoil caused by the free vortices and we now proceed to a calculation of this velocity. As before, the airfoil is replaced by a single straight vortex filament of circulation Γ which simplifies the analysis considerably. Another simplification is that the downward motion of the free vortices will be neglected as being only of secondary importance.

Fig. 160.

113. The Downward Velocity Induced by a Single Vortex Filament.—It was seen on page 206, "Fundamentals,"[1] that an element ds of a straight vortex line of circulation Γ (as shown in Fig. 160) induces at the point A a velocity,

$$dw_A = \frac{\Gamma ds \sin \varphi}{4\pi r^2},$$

or, since

$$\sin \varphi = \cos \alpha; \, s = h \tan \alpha, \, ds = \frac{h d\alpha}{\cos^2 \alpha}, \, r = \frac{h}{\cos \alpha},$$

this becomes

$$dw_A = \frac{\Gamma h d\alpha \cos \alpha}{4\pi \cos^2 \alpha \dfrac{h^2}{\cos^2 \alpha}} = \frac{\Gamma \cos \alpha}{4\pi h} d\alpha.$$

[1] See footnote, p. 3.

The direction of this velocity is perpendicular to the plane of Fig. 160.

A finite stretch of the vortex filament, the ends of which appear under the angles α_1 and α_2 as seen from A, consequently induces the velocity

$$w_A = \frac{\Gamma}{4\pi h} \int_{\alpha_1}^{\alpha_2} \cos \alpha\, d\alpha = \frac{\Gamma}{4\pi h}[\sin \alpha_2 - \sin \alpha_1]. \qquad (3)$$

For a vortex filament extending to infinity on both sides we have $\alpha_1 = -\pi/2$ and $\alpha_2 = \pi/2$, so that $\sin \alpha_2 - \sin \alpha_1 = 2$ and consequently $w_A = \Gamma/2\pi h$ (see page 207, "Fundamentals"[1]). For a vortex line extending to infinity in one direction only, the result is

$$w_A = \frac{\Gamma}{4\pi h}(1 - \sin \alpha_1),$$

which in the special case $\alpha_1 = 0$ reduces to

$$w_A = \frac{\Gamma}{4\pi h},$$

being exactly half the velocity induced by a vortex filament extending to infinity in both directions. For the practical calculation of the downward velocity induced at the airfoil by the vortex band it is assumed that this band extends to infinity back of the airplane so that the velocity at the airfoil becomes equal to half the velocity far behind the plane, where the vortex band approximately extends to infinity in both directions.

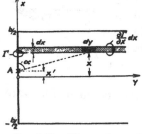

Fig. 161.—The surface of discontinuity behind a wing of given lift distribution.

114. Determination of the Induced[2] Drag for a Given Lift Distribution.—It is assumed that the lift distribution, and consequently the circulation Γ as a function of x, is known (Fig. 161). The problem is to calculate the velocity field induced by the vortex band behind the airfoil, especially the downward velocities at the wing. From this the drag can be calculated.

[1] See footnote, p. 3.

[2] This designation is used in analogy with phenomena of electromagnetic induction which are quite similar to those of hydrodynamic vortex fields. The phrase "induced drag" was first introduced by M. Munk.

The downward velocity due to the entire vortex filament emanating from B, Fig. 160, was found to be

$$w_A = \frac{\Gamma}{4\pi h}.$$

Applying this result to the strip dx of the vortex band shown in Fig. 161, the velocity at an arbitrary point x' of the bound vortex becomes equal to

$$dw(x') = \frac{1}{4\pi(x' - x)}\frac{\partial\Gamma}{\partial x}dx.$$

The direction of w is along the positive z-axis, because to the right $(x > x')$ the expression $\partial\Gamma/\partial x$ is negative, while to the left $(x < x')$ $\partial\Gamma/\partial x$ is positive.

Since the induced velocity at any point of the bound vortex (among others at the point x') is due to *all* the strips making up the vortex band, this velocity is found by integrating along the span b:

$$w(x') = \frac{1}{4\pi}\int_{-\frac{b}{2}}^{\frac{b}{2}} \frac{1}{x' - x}\frac{\partial\Gamma}{\partial x}dx.$$

Since the integral becomes indeterminate at $x = x'$ on account of the integrand becoming infinitely large, it is necessary to take the so-called "principal value" of it, defined by

$$\lim_{\epsilon \to 0}\left(\int_{-\frac{b}{2}}^{x'-\epsilon} + \int_{x'+\epsilon}^{\frac{b}{2}}\right).$$

This definition is such that the value of x' has to be approached from both sides at the same rate. Both parts of the integral tend to infinity, but their sum has a finite limit which can be checked up by calculating the velocity w at a point somewhat above or below the bound vortex instead of exactly on it. This calculation will show that the velocity remains finite and goes to a finite limit when the point is moved to the bound vortex itself.

We are now in a position to calculate the drag induced by the system of free vortices. It was seen that the various elements of the bound vortex are subjected to different downward velocities due to the free vortex band. It is assumed that each individual element of the bound-vortex filament behaves like an element of an infinitely long wing (two-dimensional flow), where the relative wind velocity is made up of the velocity V of the wing and the induced velocity w. Since in such an element the resultant force is perpendicular to the relative air velocity, it is seen from Fig. 162 that

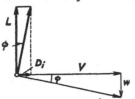

FIG. 162.—Deviation of the resulting air force due to the induced downward velocity w. Its component in the flight direction is the induced drag D_i.

$$dD = \frac{w\,dL}{V}.$$

Assuming that the lift per unit length L_1 along the span is given as a function of x', i.e., $= L_1(x')$, the total induced drag is found by integration along the span of the wing:

$$D = \frac{1}{V}\int_{-\frac{b}{2}}^{\frac{b}{2}} L_1(x')w(x')dx'.$$

Since $L_1(x') = dL/dx' = \rho\Gamma V$, the drag also can be written

$$D = \rho\int_{-\frac{b}{2}}^{\frac{b}{2}} \Gamma(x')w(x')dx',$$

or substituting into this the value for $D_1(x')$, found previously,

$$D = \frac{\rho}{4\pi}\int_{-\frac{b}{2}}^{\frac{b}{2}}\int_{-\frac{b}{2}}^{\frac{b}{2}} \Gamma(x')\frac{\partial\Gamma(x)}{\partial x}dx\frac{dx'}{x' - x}.$$

The physical meaning of the double integration (first with x' and then with x as the variable) is that first the influence of the total free vortex band on the velocity of one element of the bound vortex is determined and that further all the individual drags for the various elements have to be added up to the total induced drag of the wing. The formula gives a general method for the calculation of the induced drag if the distribution of the lift along the span is known.

There is a simple relation between the induced drag and the kinetic energy of the free vortex system, which was used by Trefftz[1] as a basis for deriving the above result by means of Green's theorem. His analysis is of considerable interest to

[1] TREFFTZ, E., The Airfoil and Propeller Theory of Prandtl (German), *Z. angew. Math. Mech.*, vol. 1, p. 206, 1921.

mathematically inclined students, but it is physically less obvious than the proof given here.

Before investigating whether these results are in accordance with the experimental facts, some remarks will be made regarding the historical development of the theory. After the formula was derived, it was attempted to find by trial some plausible functions for the lift distribution, which are simple enough to make the integration practically possible. After some time a semi-elliptic lift distribution on the span was found to lead to a simple solution, and still later it was

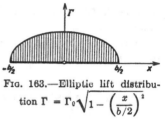

Fig. 163.—Elliptic lift distribution $\Gamma = \Gamma_0 \sqrt{1 - \left(\dfrac{x}{b/2}\right)^2}$

discovered that this solution is the most important one, since it gives a minimum of induced drag.

Designating as before by b the span and by Γ_0 the circulation in the middle of the airfoil, the elliptical lift distribution shown in Fig. 163 is expressed by

$$\Gamma = \Gamma_0 \sqrt{1 - \left(\frac{x}{b/2}\right)^2}.$$

Other lift distributions rendering the integration comparatively simple are shown in Figs. 164 and 165 and can be expressed by

Fig. 164.

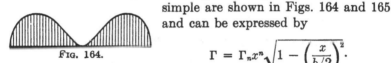

Fig. 165.

Figs. 164 and 165.—Lift distributions according to $\Gamma = \Gamma_n x^n \sqrt{1 - \left(\dfrac{x}{b/2}\right)^2}$ and linear superpositions of these.

$$\Gamma = \Gamma_n x^n \sqrt{1 - \left(\frac{x}{b/2}\right)^2}.$$

Any linear combination of terms of this sort can also be used for the purpose.

Limiting ourselves to the most important case (of elliptic lift distribution), we have

$$\frac{d\Gamma}{dx} = -\frac{\Gamma_0 x}{\dfrac{b}{2}\sqrt{\left(\dfrac{b}{2}\right)^2 - x^2}}$$

and the downward velocity at the point x' of the airfoil becomes

$$w(x') = -\frac{\Gamma_0}{4\pi} \int_{-\frac{b}{2}}^{\frac{b}{2}} \frac{x\,dx}{(x' - x)\left(\dfrac{b}{2}\right)^2 \sqrt{1 - \left(\dfrac{x}{b/2}\right)^2}}.$$

Substituting $\xi = \dfrac{x}{b/2}$, this leads to:[1]

$$w(x') = -\frac{\Gamma_0}{4\pi\frac{b}{2}} \int_{-1}^{+1} \frac{\xi d\xi}{(\xi' - \xi)\sqrt{1 - \xi^2}} = \frac{\Gamma_0}{4\pi\frac{b}{2}} \cdot \pi = \frac{\Gamma_0}{2b}.$$

It is seen therefore that the induced downward velocity w is independent of x', *i.e.*, w is constant along the entire span. It is more convenient to introduce the total lift L instead of Γ_0, and since

$$L = \rho V \int_{-\frac{b}{2}}^{\frac{b}{2}} \Gamma dx = \rho V \Gamma_0 \int_{-\frac{b}{2}}^{\frac{b}{2}} \sqrt{1 - \left(\frac{x}{b/2}\right)^2}\, dx = \rho V \Gamma_0 b \frac{\pi}{4},$$

the result becomes

$$w = \frac{2L}{\pi \rho V b^2}.$$

Since w comes out to be constant along the span, it is not necessary to perform the second integration and the induced drag immediately becomes

$$D = \frac{w}{V}L = \frac{L^2}{\pi b^2 \frac{\rho}{2} V^2}.$$

Comparing this result with the expression found on page 191,

$$D = \frac{L^2}{4S'\frac{\rho}{2}V^2},$$

it is seen that now the determination of the area S' has been accomplished:

$$S' = \frac{\pi b^2}{4}.$$

This can be conveniently memorized by noting that the area S' is equal to that of a circle on the span as diameter. It may be mentioned again that this result is true only for an elliptical lift distribution. The same result could have been obtained by pursuing the procedure started on page 195 for the two-dimen-

---

[1] The value of the integral $\displaystyle\int_{-1}^{+1} \frac{\xi d\xi}{(\xi' - \xi)\sqrt{1 - \xi^2}} = -\pi$. See Betz, footnote, p. 204.

sional flow determined by a constant velocity w_1. It will be shown later that the velocity w_1, far behind the wing, is exactly twice the velocity w at the wing itself.

The induced velocities for a given lift distribution and its consequent induced drag have now been calculated, but the question is as yet open which shape has to be given to the wing in order to obtain the desired lift distribution. In order to find an answer to this problem, the airfoil is subdivided into elemental strips of width dx, each having a definite circulation determined by the known lift distribution. The question arises as to which shape each element has to have in order that the given circulation correspond to it if the element is considered to be part of an infinitely long wing. Besides depending on the shape of the profile, the circulation also is affected by the chord and by the angle of attack so that it is clear that the problem of determining the wing shape for a given lift distribution is an indeterminate one. Various shapes corresponding to the same lift distribution can be obtained by varying either the profile or the angle of attack, or the chord along the length of the span. The most practical case from a structural standpoint is to make the profiles of the various elements geometrically similar and their angles of attack equal along the span. With this restriction the question of the wing shape can be solved by making the chord at each point proportional to the lift. An elliptical lift distribution therefore can be realized by making a wing consist of two semi-ellipses as shown in Fig. 166. By this special choice

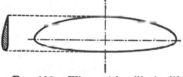

Fig. 166.—Wing with elliptic lift distribution; the shape of the wing consists of two semi-ellipses.

the additional advantage is obtained that the centers of pressure of the individual profiles are all on a straight line so that this particular wing can be approximated very well by a straight vortex filament. In case the bound vortex should be curved somewhat in the xy-plane, the angles of attack would be changed by the induced velocities of the individual elements of the bound vortex on each other. This would be very difficult to follow analytically.

115. Minimum of the Induced Drag; the Lift Distribution of an Airfoil of Given Shape and Angle of Attack.—The so-called "second problem of airfoil theory" consists of finding the lift distribution for a given total lift and a given span such that the induced drag is a minimum. In mathematical language the problem is therefore:

$$D = \int_{-\frac{b}{2}}^{\frac{b}{2}} \Gamma(x)w(x)dx = \text{minimum}$$

when

$$L = \rho V \int_{-\frac{b}{2}}^{\frac{b}{2}} \Gamma(x)dx$$

is given, and

$$w(x) = \frac{1}{4\pi} \int_{-\frac{b}{2}}^{\frac{b}{2}} \frac{\frac{\partial \Gamma}{\partial x}dx}{x' - x}.$$

This problem has been solved in its most general form by Munk[1] for monoplanes as well as for multiplanes with the result that for a monoplane the minimum is obtained for a velocity w which is constant along the span. As was mentioned before, the elliptic lift distribution therefore is the one which gives a minimum induced drag for a given span and a given total lift. A simpler proof of this theorem given later by Betz is discussed in Art. 120.

The minimum in the drag for the elliptic lift distribution is a very flat one, however, so that the drag does not increase much for lift distributions different from the elliptic one. For instance, a rectangular wing of aspect ratio 5 has only 4 per cent greater induced drag than the corresponding elliptic wing.

The "third problem of airfoil theory" consists of finding the lift distribution for an airfoil of given shape and given angle of attack. Naturally this problem was the first to present itself, but being the most complicated one it was last to be solved in 1919 by Betz.[2] The problem leads to a disagreeable integro-differential equation which has been solved for the case of the rectangular wing of constant profile and constant angle of attack

[1] MUNK, M., Isoperimetric Problems in the Theory of Flight (German), Dissertation, Göttingen, 1919.

[2] BETZ, A., On Airfoil Theory with Special Consideration of Rectangular Wings (German), Dissertation, Göttingen, 1919.

in the form of a series of powers of a parameter P proportional to the aspect ratio

$$P = \frac{2}{\pi} \frac{b}{c}.$$

The calculation is simpler for small aspect ratios than for large ones. Another approximate solution which can be applied for large aspect ratios has been found by Trefftz.[1]

The result of these calculations is that for very small aspect ratios the lift distribution is practically elliptical; for larger aspect ratios the distribution becomes flatter, while for the case

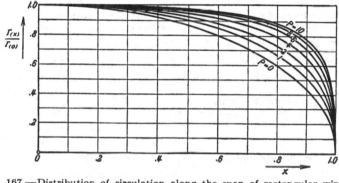

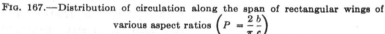

Fig. 167.—Distribution of circulation along the span of rectangular wings of various aspect ratios $\left(P = \dfrac{2}{\pi} \dfrac{b}{c} \right)$

of an infinitely long wing the rectangular distribution results (Fig. 167).

The induced downward velocity w becomes smaller at the middle of the wing and greater at the tips when the aspect ratio increases. As was to be expected, the induced downward velocity and induced drag become zero when the aspect ratio (and consequently the parameter P) go to infinity (two-dimensional problem). According to Trefftz's calculations, a wing extending to infinity to one side only has a finite downward velocity at its one end and also a finite induced drag. According to the calculations of Betz, an approximate formula for the relation between the induced drag and the aspect ratio in the region $P = 1$ to $P = 10$ is

$$\frac{D}{\dfrac{2L^2}{\pi \rho V^2 b^2}} = \frac{D}{D_{\min}} = 0.99 + 0.015P \quad \text{[see Fig. 168].}$$

[1] See footnote, p. 200.

The induced drag is not distributed uniformly along the span but is concentrated at the ends, which tendency is very pronounced for large aspect ratios. Figure 169 shows the lift, the induced velocity, and the induced drag of a very long wing.

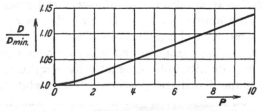

Fig. 168.—Ratio of drag of wing with rectangular lift distribution and one of elliptic distribution ($D_{min.}$) for various aspect ratios ($P = 2b/\pi c$).

116. Conversion Formulas.—Concerning the relation between the calculated induced drag and the experimental results, it can be stated with certainty that the experimental drag cannot

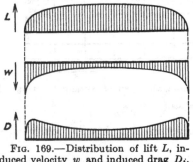

Fig. 169.—Distribution of lift L, induced velocity w and induced drag D_i, along rectangular wing of large aspect ratio.

be smaller than the calculated one but must be necessarily greater. There are two parts of the total drag which have not been taken account of in our calculations thus far, namely, the skin friction and the small eddy resistance due to the fact that the streams from above and below the profile do not join smoothly. These two partial drags together have been called the "profile drag."

Taking into consideration the fact that the total drag is the sum of the calculated induced drag and the profile drag, the agreement between theory and experiment is very satisfactory. For the induced drag coefficient C_{D_i} we found on page 202

$$C_{D_i} = \frac{C_L{}^2 S}{4S'}$$

or, with

$$S' = \frac{\pi b^2}{4},$$

this becomes:

$$C_{D_i} = \frac{C_L{}^2 S}{\pi b^2}.$$

For the rectangular wing, we have specifically $S = bc =$ span × chord, so that

$$C_{D_i} = \frac{C_L{}^2 c}{\pi b}.$$

Plotting this relation graphically in Fig. 170, it is seen that the induced drag becomes a parabola with a curvature depending on the aspect ratio. Plotting into the same diagram the polar curve of a good wing of the same aspect ratio shows that the total drag consists for its larger part of induced drag, especially for large angles of attack.

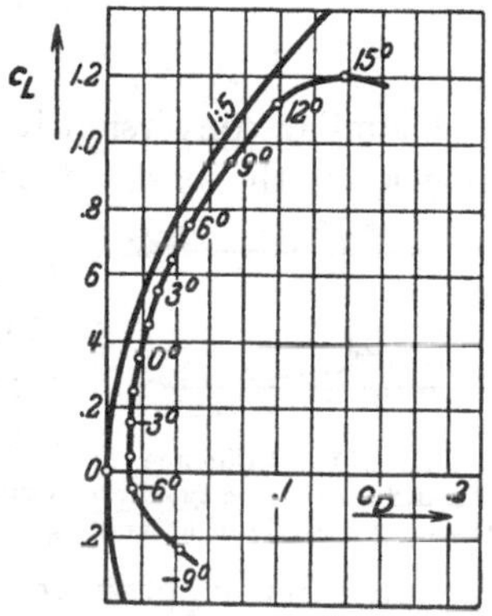

On account of a fortunate coincidence, which was not anticipated, it was possible to reduce the results of airfoil theory to a very useful form. The polar diagrams for a number of airfoils of the same profile but of different aspect ratio were all plotted on the same curve sheet and it was seen that the difference between the calculated induced-drag coefficient and the measured

Fig. 170.—Induced drag parabola for aspect ratio 5 together with experimental curve.

total drag coefficient was about equal in all cases. From this it was concluded that the profile-drag coefficient was practically independent of the aspect ratio, so that the possibility presented itself to convert polar curves from one aspect ratio to another.

The problem therefore is to calculate from a given polar curve 1, for a given aspect ratio $b_1{}^2/S_1$ another polar curve 2 for the same profile but of a different aspect ratio $b_2{}^2/S_2$. In other words, it is necessary to calculate for various values of C_L the values C_{D_2} from the values C_{D_1}.

First, we split up the total drag coefficient into its induced and profile parts, *i.e.*,

$$C_D = C_{D_i} + C_{D_p},$$

where C_{D_p} is a function of C_L. For a given profile with the aspect ratio $b_1{}^2/S_1$ we have

$$C_{D_1} = \frac{C_L{}^2}{\pi} \frac{S_1}{b_1{}^2} + C_{D_p}.$$

For the same profile with the different aspect ratio $b_2{}^2/S_2$ we have consequently

$$C_{D_2} = \frac{C_L{}^2}{\pi}\frac{S_2}{b_2{}^2} + C_{D_p}$$

so that the conversion formula becomes

$$C_{D_2} = C_{D_1} + \frac{C_L{}^2}{\pi}\left(\frac{S_2}{b_2{}^2} - \frac{S_1}{b_1{}^2}\right)$$

A similar conversion formula can be derived for the angle of attack on the basis of an elliptical lift distribution. We first

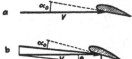

consider an element of an infinite airfoil (two-dimensional flow), as shown in Fig. 171a. Assuming that the element now is a part of a wing of finite length, we know that owing to the free vortex band the velocity V at the wing is subjected to an induced downward velocity component

FIG. 171.—Influence of the induced velocity w on the actual angle of attack.

w. In order to obtain geometrical similarity between the element of the finite wing and the same element when considered as a part of an infinite wing, it is necessary to turn the element through an angle φ determined by

$$\tan \varphi = \frac{w}{V} = \frac{C_L}{\pi}\frac{S}{b^2}.$$

The angle of attack which the element of the finite wing would have if it were an element of an infinite wing for the same lift consequently is (see Fig. 171b)

$$\alpha_0 = \alpha - \varphi.$$

Considering that w is small with respect to V and that consequently φ can be set equal to $\tan \varphi$, two airfoils of the same profile but different aspect ratios are in the same condition if their actual angles of attack α_0 are equal for the same lift coefficient, *i.e.*,

$$\alpha_0 = \alpha_1 - \frac{C_L}{\pi}\frac{S_1}{b_1{}^2} = \alpha_2 - \frac{C_L}{\pi}\frac{S_2}{b_2{}^2}$$

so that

$$\alpha_2 = \alpha_1 + \frac{C_L}{\pi}\left(\frac{S_2}{b_2{}^2} - \frac{S_1}{b_1{}^2}\right).$$

This is the conversion formula for angles of attack. Both conversion formulas have been derived only for airfoils where the lift distribution is elliptic along the span. This is not serious, however, since the drag is a minimum for the elliptic distribution and consequently varies only little even with drastic departures from the elliptic loading. Moreover the lift distribution for rectangular wings is not very much different from the elliptic one (Fig. 167). This makes the conversion formulas applicable with sufficient accuracy to almost any type of wing.

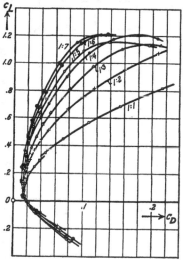

Fig. 172.—Polar diagrams for wings of the same profile and various aspect ratio (from 1 to 7).

In Figs. 172 to 175, an example is given of the use of these relations. Figures 172 and 173 show the polar diagrams and the

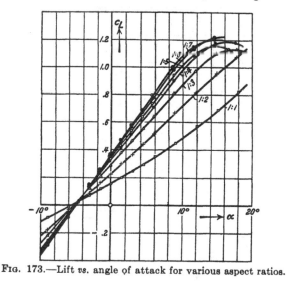

Fig. 173.—Lift *vs.* angle of attack for various aspect ratios.

relation between the lift coefficient and the angle of attack for seven wings of aspect ratios ranging from 1 to 7. In Figs. 174

and 175 these two diagrams have been converted to an aspect

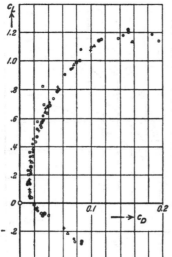

ratio 5 by means of the conversion formulas. It is seen that the various experimental points lie on a smooth curve, with the exception of a few points for the wing of aspect ratio 1, and this is not surprising since the whole theory is based on the concept of a bound straight vortex filament to which a square wing cannot be approximated with sufficient accuracy.

117. Mutual Influence of Bound Vortex Systems. The Unstaggered Biplane.—It has been proved before that the free vortex band in the wake is responsible for a downward induced velocity at the wing and consequently for an induced drag. The influence exists

Fig. 174.—Figure 172 replotted for aspect ratio 5 by conversion formula.

between the free vortex system caused by a bound vortex

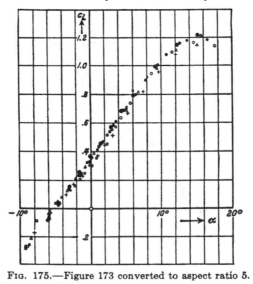

Fig. 175.—Figure 173 converted to aspect ratio 5.

and that bound vortex *itself* and therefore can be called a case of "*self*-induction." In the case of biplanes or multiplanes, there also may be an influence of the free vortex band of one wing on the bound vortex of another wing which might be called *"mutual* induction." A theory will now be developed, whereby it becomes possible to calculate the induced drag of a biplane or a multiplane from the wind-tunnel data of a single wing.

In principle the action of mutual induction in a biplane consists of a downward induced velocity at wing 1 due to the free vortex band behind wing 2 and *vice versa.*

Each wing therefore has a self-induced drag due to its own free vortex band and a mutually induced drag due to the vortex

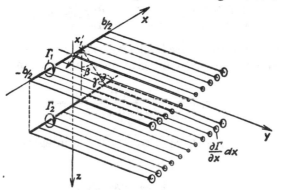

Fig. 176.—Surfaces of discontinuity with unstaggered biplane.

band of the other wing. The total induced drag of a biplane therefore consists of four terms,

$$D = D_{11} + D_{12} + D_{21} + D_{22},$$

where D_{11} denotes the self-induced drag of wing 1; D_{12} is the drag of wing 1 due to the influence of wing 2; in the same manner D_{21} is the drag of wing 2 due to the influence of wing 1 and D_{22} is the self-induced drag of wing 2.

Besides inducing a downward component of velocity w, one of the wings of a biplane induces also a horizontal velocity v, causing an increase or a decrease in the relative wing velocity. The change in the drag due to this effect, however, is small of the second order, so that in the following calculation the influence of the horizontal component v will be neglected.

First the case of an unstaggered biplane will be discussed, *i.e.,* of a biplane where the two wings are perpendicularly above one another. As before, the wings are replaced by straight, parallel, bound vortex filaments as shown in Fig. 176. In this case neither wing induces a vertical velocity at the other one so that the downward induced velocity at either wing is due only to the free vortex bands.   First, the induced velocity will be calculated at a point x_1' of wing 1 caused by the free vortices of wing 2.   The vortex strength of an element dx_2 of this band is $\partial\Gamma_2/\partial x_2 dx_2$ so that according to page 198 this velocity calculates to

$$\frac{1}{4\pi}\frac{\partial\Gamma_2}{\partial x_2}\frac{dx_2}{a}$$

with a vertical component

$$\frac{1}{4\pi}\frac{\partial\Gamma_2}{\partial x_2}\frac{dx_2}{a}\cos\gamma = -\frac{1}{4\pi}\frac{\partial\Gamma_2}{\partial x_2}\frac{dx_2}{a}\sin\beta.$$

The total vertical induced velocity at x_1' due to the entire free band of wing 2 is found by integration along the span:

$$w(x_1') = -\frac{1}{4\pi}\int_{-\frac{b}{2}}^{\frac{b}{2}}\frac{\partial\Gamma_2}{\partial x_2}\frac{\sin\beta}{a}dx_2.$$

Considering that $\Gamma = 0$ for $x_2 = \pm b/2$ and integrating by parts, we obtain

$$w(x_1') = \frac{1}{4\pi}\int_{-\frac{b}{2}}^{\frac{b}{2}}\Gamma_2\frac{\partial}{\partial x_2}\left(\frac{\sin\beta}{a}\right)dx_2.$$

By means of the relations

$$\frac{\partial}{\partial x}\left(\frac{\sin\beta}{a}\right) = \frac{\partial}{\partial x}\left(\frac{x_1'-x_1}{a^2}\right) = \frac{a^2-2(x_1'-x_1)^2}{a^4}$$

$$= \frac{1-2\sin^2\beta}{a^2} = \frac{\cos 2\beta}{a^2}$$

this can be simplified to

$$w(x_1') = \frac{1}{4\pi}\int_{-\frac{b}{2}}^{\frac{b}{2}}\Gamma_2\frac{\cos 2\beta}{a^2}dx_2.$$

Now the expression for the induced drag D_{12} can be found by means of the relations of page 200 and is

$$D_{12} = \rho\int_{-\frac{b}{2}}^{\frac{b}{2}}\Gamma_1 w(x_1')dx_1$$

or, after substituting the value for $w(x_1')$,

$$D_{12} = \frac{\rho}{4\pi} \int_{-\frac{b}{2}}^{\frac{b}{2}} \int_{-\frac{b}{2}}^{\frac{b}{2}} \Gamma_1 \Gamma_2 \frac{\cos 2\beta}{a^2} dx_1 dx_2. \tag{1}$$

From the symmetrical structure of this integral it is seen that the same result would have been obtained for D_{21}, so that

$$D_{12} = D_{21}.$$

This theorem, which was derived in a different manner by Munk,[1] states that for an unstaggered biplane the two mutually induced drags are equal.

Although this relation was derived on the basis of straight bound vortices, it is valid also for curved bound vortex filaments provided these filaments lie in a plane vertical to the direction of flight. In this case, Eq. (1) has to be changed in so far as $\cos (\beta_1 + \beta_2)$ has to be substituted for $\cos 2\beta$, where β_1 and β_2 are the angles of the connecting line a with the elements of the bound vortex filament; further, $dx_1 dx_2$ has to be replaced by $ds_1 ds_2$.

The mutually induced drag is always positive for unstaggered biplanes of the ordinary type. With tandem biplanes, where the two wings are beside each other in the same line, the mutual influence is different in so far as each wing is in the field of an upward current of air caused by the other wing and consequently the mutually induced drag is negative. In such a case the total drag of the two wings is less than the sum of the drags of each wing by itself.

118. The Staggered Biplane.—For the staggered biplane, the bound vortex of the one wing induces a vertical velocity at the other wing, so that in addition to the influence of the free vortex band the influence of the bound vortex has to be considered. For a point x_1' of wing 1, the vertical induced velocity due to wing 2 and its vortex band therefore consists of the following two contributions:

1. The velocity w_1 due to the free vortex band of wing 2,
2. The velocity w_2 due to the bound vortex of wing 2 itself.

A strip dx of the free vortex band has the strength $\partial \Gamma_2 / \partial x_2 dx_2$, and the induced velocity due to it (see page 198 and Fig. 177) is

[1] See footnote, p. 204.

$$\frac{1}{4\pi a}\frac{\partial \Gamma_2}{\partial x_2}dx_2 \sin \alpha \Big|_{\alpha}^{\pi/2} = \frac{1}{4\pi a}\frac{\partial \Gamma_2}{\partial x_2}dx_2(1 - \sin \alpha),$$

of which the vertical component is

$$-\frac{1}{4\pi a}\frac{\partial \Gamma_2}{\partial x_2}dx_2(1 - \sin \alpha)\sin \beta.$$

The total induced velocity at x_1' due to the entire free vortex band of wing 2, therefore, is

$$w_1(x_1') = -\frac{1}{4\pi}\int_{\frac{b}{2}}^{\frac{b}{2}}\frac{\partial \Gamma_2}{\partial x_2}\frac{1 - \sin \alpha}{a}\sin \beta dx_2$$

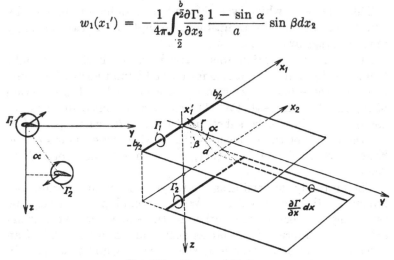

FIG. 177.—Staggered biplane.

or considering that $\sin \alpha = y/r$ and $\sin \beta = -(x_1' - x_1)/a$,

$$w_1(x_1') = \frac{1}{4\pi}\int_{-\frac{b}{2}}^{\frac{b}{2}}\frac{\partial \Gamma_2}{\partial x_2}\frac{1 - \dfrac{y}{r}}{a}\frac{x_1' - x_1}{a}dx_2.$$

Since Γ becomes zero for $-b/2$ and $+b/2$ the expression can be integrated by parts in the same manner as was done on page 212,

$$w_1(x_1') = \frac{1}{4\pi}\int_{-\frac{b}{2}}^{\frac{b}{2}}\Gamma_2\frac{\partial}{\partial x_2}\left[\frac{x_1' - x_1}{a^2}\left(1 - \frac{y}{r}\right)\right]dx_2.$$

Performing the differentiation and remembering that $r = \sqrt{a^2 + y^2}$ and $a = \sqrt{(x_1' - x_1)^2 + z^2}$, this becomes

$$w_1(x_1') = \frac{1}{4\pi}\int_{-\frac{b}{2}}^{\frac{b}{2}}\Gamma_2\left[\frac{a^2 - 2(x_1' - x_1)^2}{a^4}\left(1 - \frac{y}{r}\right) - \frac{y(x_1' - x_1)^2}{a^2r^3}\right]dx_2.$$

The second contribution to the vertical induced velocity due to the bound vortex of the wing 2 at the point x_1' becomes

$$w_2(x_1') = -\frac{1}{4\pi}\int_{-\frac{b}{2}}^{\frac{b}{2}}\frac{\Gamma_2 \sin \alpha}{r^2}dx_2 = -\frac{1}{4\pi}\int_{-\frac{b}{2}}^{\frac{b}{2}}\Gamma_2\frac{y}{r^3}dx_2,$$

if the lower wing 2 is staggered behind the upper wing 1, as shown in Fig. 177. It is seen that this vertical component is directed upward so that the wing 2 causes a decrease in the drag of wing 1. In case the stagger had been reversed and the lower wing had been placed in front of the upper one, this effect would have been reversed and the induced drag of wing 1 would have been increased.

The total induced downward velocity at the point x_1' therefore becomes

$$w_1 + w_2 = w(x_1') =$$

$$\frac{1}{4\pi}\int_{-\frac{b}{2}}^{\frac{b}{2}}\Gamma_2\left[\frac{a^2 - 2(x_1' - x_1)^2}{a^4}\left(1 - \frac{y}{r}\right) - \frac{y(x_1' - x_1)^2}{a^2r^3} - \frac{y}{r^3}\right]dx_2$$

or since

$$\frac{a^2 - 2(x_1' - x_1)^2}{a^4} = \frac{\cos 2\beta}{a^2}, \qquad \frac{y}{r} = \sin \alpha,$$

and

$$\frac{y(x_1' - x_1)^2}{a^2r^3} - \frac{y}{r^3} = \frac{1}{r^2}\left[\frac{y}{r}\left\{\frac{(x_1' - x_1)^2}{a^2} - 1\right\}\right]$$

$$= \frac{1}{r^2}\left[\frac{y}{r}\left(\frac{a^2 - z^2}{a^2} - 1\right)\right] = -\frac{1}{r^2}\frac{y}{r}\left(\frac{z}{a}\right)^2 = -\frac{\sin \alpha \cos^2 \beta}{r^2},$$

this becomes

$$w(x_1') = \frac{1}{4\pi}\int_{-\frac{b}{2}}^{\frac{b}{2}}\Gamma_2\left[\frac{\cos 2\beta}{a^2}(1 - \sin \alpha) - \frac{\sin \alpha \cos^2 \beta}{r^2}\right]dx_2.$$

Since, according to page 212, the drag of wing 1 induced by wing 2 is equal to

$$D_{12} = \rho \int_{-\frac{b}{2}}^{\frac{b}{2}} \Gamma_1 w(x_1) dx_1,$$

this becomes on substitution of the above value for $w(x_1)$:

$$D_{12} = \frac{\rho}{4\pi} \int_{-\frac{b}{2}}^{\frac{b}{2}} \int_{-\frac{b}{2}}^{\frac{b}{2}} \Gamma_1 \Gamma_2 \left[\frac{\cos 2\beta}{a^2}(1 - \sin \alpha) - \frac{\sin \alpha \cos^2 \beta}{r^2} \right] dx_1 dx_2.$$

As is seen in Fig. 177, the drag of wing 2, due to the effect of wing 1, can be obtained from the previous formula by putting $\alpha + \pi$ and $\beta + \pi$ instead of α and β. Consequently

$$D_{12} = \frac{\rho}{4\pi} \int_{-\frac{b}{2}}^{\frac{b}{2}} \int_{-\frac{b}{2}}^{\frac{b}{2}} \Gamma_1 \Gamma_2 \left[\frac{\cos 2\beta}{a^2}(1 + \sin \alpha) + \frac{\sin \alpha \cos^2 \beta}{r^2} \right] dx_1 dx_2.$$

For $\alpha = 0$ it is seen that the two integrals become equal $(D_{12} = D_{21})$, which is the result for the unstaggered biplane obtained before. It was shown first by Munk that the sum $D_{12} + D_{21}$ is independent of the amount of stagger (stagger theorem).

For the general case of bound vortex filaments which are not parallel, the final result becomes

$$D_{12} + D_{21} = \frac{\rho}{2\pi} \int_{-\frac{b}{2}}^{\frac{b}{2}} \int_{-\frac{b}{2}}^{\frac{b}{2}} \Gamma_1 \Gamma_2 \frac{\cos (\beta_1 + \beta_2)}{a^2} ds_1 ds_2,$$

where also the independence of the angle of stagger α is apparent.

It is important to note that this theorem of the independence of the total mutual induced drag from the amount of stagger is true only if the lift distribution of the two wings is not changed, and this is possible only by properly changing the angles of attack of the various wing elements. An alteration in the stagger without changing the angles of attack would result in a change of the effective angles of attack and consequently in a change of the lifts of the two wings. The geometrical angles of attack have to be corrected in such a manner that for a change in the stagger the effective angles of attack remain the same.

119. The Total Induced Drag of Biplanes.—It was seen on page 202 that the self-induced drags of the two wings of a biplane are expressed by

$$D_{11} = \frac{L_1{}^2}{\pi\frac{\rho}{2}V^2 b_1{}^2},$$

$$D_{22} = \frac{L_2{}^2}{\pi\frac{\rho}{2}V^2 b_2{}^2},$$

if the lift distribution is elliptical. L_1 is the lift on the first wing and L_2 that on the second wing. Analogously, the mutual induced drag D_{12} or D_{21} can be represented by

$$\sigma\frac{L_1 L_2}{\pi\frac{\rho}{2}V^2 b_1 b_2},$$

where the coefficient σ depends on the ratio b_1/b_2 of the spans and on $2z/(b_1 + b_2)$, where z is the distance between the wings in a direction vertical to the direction of flight. For an elliptical lift distribution, the value of σ has been calculated on the assumption that the centers of the two straight wings are in the same plane of symmetry. Figure 178 shows the relation between σ and $2z/(b_1 + b_2)$ for three different values of b_2/b_1. Using this figure, the total induced drag of the biplane can be calculated from

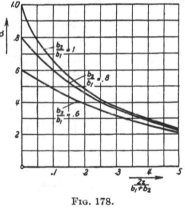

Fig. 178.

$$D = D_{11} + 2D_{12} + D_{22} = \frac{1}{4\pi\frac{\rho}{2}V^2}\left(\frac{L_1{}^2}{b_1{}^2} + 2\sigma\frac{L_1 L_2}{b_1 b_2} + \frac{L_2{}^2}{b_2{}^2}\right), \quad (1)$$

if the lift distribution between the two wings is known.

It is of interest to know the distribution of the total lift between the two wings for which the total induced drag becomes a minimum. A simple calculation shows that this is the case if

$$\frac{L_2}{L_1} = \frac{\dfrac{b_2}{b_1} - \sigma}{\dfrac{b_1}{b_2} - \sigma}.$$

In order to obtain this result, we put $L_1 = \lambda L$ and consequently $L_2 = (1 - \lambda)L$ and then determine the value of λ for which the parenthesis of Eq. (1) becomes a minimum. The value of this minimum then is found to be

$$D_{\min} = \frac{(L_1 + L_2)^2}{\pi \frac{\rho}{2} V^2 b_1{}^2} \frac{1 - \sigma^2}{1 - 2\sigma \frac{b_2}{b_1} + \left(\frac{b_2}{b_1}\right)^2} = \frac{(L_1 + L_2)^2}{\pi \frac{\rho}{2} V^2 b_1{}^2} \kappa. \qquad (2)$$

Since the factor

$$\frac{(L_1 + L_2)^2}{\pi \frac{\rho}{2} V^2 b_1{}^2}$$

represents the drag of a monoplane of span b_1 with a lift $L_1 + L_2$, and since the second factor κ is always smaller than unity ($\sigma <$

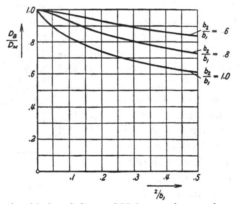

Fig. 179.—Ratio of induced drags of biplane and monoplane of the same lift and span b_1 as a function of the height h/b_1 for various values of b_1/b_2. The induced drag is a minimum for $b_1 = b_2$.

b_2/b_1), it is seen that the total induced drag of a biplane D_B is smaller than that of a monoplane D_M of the same span b_1 and of the same total lift. Figure 179 shows the relation between D_B/D_M and z/b_1 for a number of values b_2/b_1. It is seen that the drag of the biplane decreases rapidly with increasing z and b_2/b_1. Therefore the most advantageous arrangement of a biplane is the one where both wings have the same span, *i.e.*, $b_1 = b_2$.

However, the advantage of the biplane as compared to the monoplane is not so important as these results would indicate.

A relatively small increase in the span of the monoplane makes it possible to decrease its induced drag to the same value as that of the biplane. The factor by which the span of a monoplane has to be multiplied in order to give it the same induced drag as a biplane of the same lift is denoted by x. We have according to Eq. (2)

$$\frac{\kappa}{b_1{}^2} = \frac{1}{(xb_1)^2}$$

or

$$x = \frac{1}{\sqrt{\kappa}}.$$

Figure 180 shows how the span of a monoplane has to be increased to obtain the same induced drag as the biplane $b_1 = b_2$.

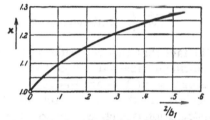

Fig. 180.—The ordinate x is the factor with which the span b_1 of a biplane has to be multiplied in order to get a monoplane of the same lift and the same induced drag. The biplane has two wings of equal span $b_1 = b_2$. The abscissas are the height-span ratio of the biplane.

For instance, a biplane with a span of $b = 30$ ft and a vertical distance of $z = 6$ ft has the same induced drag as a monoplane of 34.8-ft span, where both planes have the same lift.

120. Minimum Theorem for Multiplanes.—After having solved the problem of minimum induced drag of a biplane with the lift distribution between the two wings as the variable, we proceed to the more general problem of determining the lift distribution over each individual wing of a biplane or multiplane of given dimensions, required to make the induced drag a minimum. This problem was solved first by Munk while Betz gave a simpler proof for it later. The result was that for either a biplane or multiplane the drag becomes a minimum when the lift distribution is such as to cause a constant downward induced velocity at both wings.

The proof given by Betz is based on Munk's theorem of stagger. The original airfoil system and the variation given to it are considered as two separate systems, the variation consisting of a very small wing system. According to the stagger theorem it is permissible to shift the variation wing system far back of the plane. In that case the downward induced velocity at the main wings due to the small variation system is negligible, and the only change in the induced drag due to variation is caused by the effect of the vortex band on the variation wing system.

If the correct lift distribution for minimum induced drag exists to start with, any additional variation system which does not change the lift will not change the drag.

Thus adding to an arbitrary location dx of one of the wings the additional lift δL_1, and simultaneously adding to another spot of the same or of another wing the lift $\delta L_2 = -\delta L_1$, the variations in the induced drag become

$$\delta L_1 \frac{w_1}{V} \quad \text{and} \quad \delta L_2 \frac{w_2}{V},$$

on account of the fact that this variation in the lift is equivalent to very small additional wings far behind the actual wing system. In these expressions, w_1 and w_2 are the vertical velocities at the corresponding points in the free vortex band. In the case where the lift distribution is such as to make the total drag a minimum, it is clear that the variation in the lift distribution must be zero or

$$\delta L_1 \frac{w_1}{V} + \delta L_2 \frac{w_2}{V} = 0.$$

Considering that $\delta L_1 = -\delta L_2$ it follows that

$$w_1 = w_2.$$

Since the velocities at the unstaggered wing system are exactly half those in the vortex band far behind the plane (see Art. 117), it follows that the velocities at the wings are also equal. The locations at which the variation of the lift was made are entirely arbitrary and it can therefore be concluded that the induced velocities are equal not only at the two points just chosen but have to be equal everywhere. This completes the proof of the theorem that for a given lift the drag becomes a minimum when the induced downward velocities at both wings are equal and constant along the span. In connection with our previous discussions (page 192), where the flow round an airfoil was compared to the impulsive downward acceleration of the entire flight path of the wing by means of a "board," it is now seen that in the case of minimum induced drag the system of "boards" representing the biplane or multiplane is a rigid one. In case the most advantageous induced velocities had not been constant, this system of "boards" should have been made flexible.

It is now possible to find a plausible and simple interpretation for the general drag formulas in the case of minimum drag. Let w_1 be the final velocity of our "board" system after the acceleration; then the downward velocity at the wing is $w = w_1/2$ and the induced drag is

$$D = L\frac{w}{V} = \frac{Lw_1}{2V}.$$

Expressing w_1 by means of the momentum theorem,

$$L = \rho S' V w_1,$$

we find, as before, that

$$D = \frac{L^2}{2\rho S' V^2}.$$

In this expression the area S' is to be put equal to $\sum \int (\varphi_a - \varphi_b) dx$, where the sum covers all individual wings. Remembering

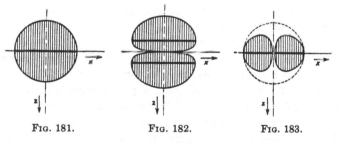

Fig. 181. Fig. 182. Fig. 183.

(see page 202) that the entire mass of air $\rho S' V$ is given the downward velocity w_1 and that the rest of the air is not affected, the equation states that the lift is equal to the momentum given to this mass of air. The work done by the drag is equal to the kinetic energy imparted by the wing to the air.

It was seen before that for the monoplane the area S' is a circle with a diameter equal to the span (Fig. 181). Grammel and Pohlhausen have calculated the corresponding areas S' for the usual biplane and for the tandem biplane by means of elliptic integrals (Figs. 182 and 183). The relation

$$D = \frac{L^2}{2\rho V^2 \cdot S'}$$

shows that the induced drag for the same lift is smaller when S' is larger. If the distance between the two wings of a biplane becomes very large, S' reduces to two circles on the span as a diameter, i.e., the biplane acts as two separate monoplanes. Such a biplane has therefore half the induced drag of a monoplane of the same lift. On the other hand, if the two wings of the biplane get closer and closer together the figure of S' finally becomes identical with the circle of the monoplane and consequently the induced drag of the biplane becomes equal to that of the monoplane.

121. The Influence of Walls and of Free Boundaries.—Airfoil theory has yielded yet another result, which is of importance for the interpretation of the experiments on models in wind tunnels either of the solid-wall or of the free-jet type.

From the tests in a free jet or in a closed tunnel it is intended to draw conclusions regarding the behavior of the test body in an atmosphere of infinite extent. The differences between the wind-tunnel stream and the free atmosphere lie in the boundary conditions: on the solid walls of a wind tunnel the normal component of the velocity must be zero, while for a free jet the pressure at the boundary is constant and equal to the pressure of the surrounding air. With the usual experiments, these deviations at the boundary of the jet cause certain changes in the flow around the model as well as in its induced drag, and it is often necessary to take these boundary effects into consideration.

We begin by assuming an infinite atmosphere and the corresponding velocity field around the wing under test. Then the jet is cut out of this infinite atmosphere and it is seen that there are lateral velocity components and pressures at the boundary of the jet thus cut out. In order to obtain the actual flow in the wind tunnel, it is necessary to superpose another flow, which has no singularities in the inside and has lateral velocity components (or pressure variations) at the surface of the jet equal and opposite to the ones of the first flow. This superposition gives the actual flow in the wind tunnel, and the action of the secondary superposed velocity field on the wing is just equal to the correction we are looking for.

Since this secondary velocity field is a potential field (having no discontinuities), it is only necessary to find the expression for its potential. This is the so-called "second problem of potential theory," where the potential function has to be determined inside a closed region when its derivatives are known on the boundary of the region.

A similar process leads to the solution of the problem for a free jet. The boundary condition here is that the pressure is constant on the surface of the jet. Denoting by V the velocity of the main flow and by u, v, w the components of the flow induced by the airfoil under test, Bernoulli's equation gives

$$p + \frac{\rho}{2}[u^2 + (V + v)^2 + w^2] = p_0 + \frac{\rho}{2}V^2,$$

and applied to the free surface of the jet, where $p = p_0$,

$$u^2 + v^2 + w^2 + 2Vv = 0.$$

Assuming the induced velocities so small that their squares can be neglected, only the last term $2Vv$ is of importance and the boundary condition for the jet therefore becomes

$$v = 0.$$

On the further assumption that the lift of the airfoil, and consequently the deviation in the direction of the jet, is small, the problem is simplified by taking $v = 0$ on the original undisturbed jet instead of on the actual deflected one. Therefore it is seen that the boundary condition for a stream between solid walls is such that the normal velocity component is zero while for a free jet the tangential component v has to be zero.

In the same manner as in the closed wind tunnel, the actual flow in a free jet is obtained by the superposition of the flow cut out from an infinite atmosphere and a secondary flow which has the velocities $-v$ at the surface of the jet. If Φ is the potential of this velocity field, the boundary condition on the surface of the jet is $\partial\Phi/\partial y = -v$. An integration along the generators of the cylinder then gives

$$\Phi(y) = -\int_{-\infty}^{y} v\,dy.$$

The lower limit of this integration expresses the fact that at a great distance from the test wing ($y = -\infty$) the potential of the secondary flow Φ is zero. Thus the condition that the secondary velocity field has the prescribed values $-v$ at the jet boundary is equivalent to the condition that the values of the potential are given on the boundary and the problem therefore is that of finding a function without singularities inside the jet while the values of this function are given on the boundary of the jet. This problem is known as the "first problem of potential theory."

122. Calculation of the Influence for a Circular Cross Section. The problem for the solid walled channel as well as that for the free jet admits of the easiest solution for the circular cross section. The problem reduces to finding the action of an "inverted" airfoil, *i.e.*, of a body found from the original airfoil by a mirroring process involving reciprocal radii. The circulation around this inverted wing has to be taken in the same sense as around the actual wing for the case of a solid walled channel and with the opposite sign for a free jet. The calculations have been carried out in detail for a straight monoplane in the middle of the jet, an elliptical lift distribution being assumed. Letting the span be b, the diameter of the jet be d, and $\xi = 2x/d$, where x is the distance from the center of the wing, it is found that

$$w'(\xi) = \frac{L}{\pi d^2 \rho V}\left(1 + \frac{3}{4}\xi^2 + \frac{5}{8}\xi^4 + \frac{35}{128}\xi^6 + \cdots\right).$$

The added drag due to this velocity consequently becomes

$$D' = \frac{L^2}{\pi d^2 \rho V^2}\left[1 + \frac{3}{16}\left(\frac{b}{d}\right)^4 + \frac{5}{64}\left(\frac{b}{d}\right)^8 + \cdots\right].$$

This expression for the additional drag is exactly true for straight monoplanes with an elliptical lift distribution; however, it is also valid with a good approximation for all usual wing systems of which the dimensions relative to the diameter of the jet are not too great. The expression for the induced drag was found on page 202, namely,

$$D = \frac{L^2}{4\frac{\rho}{2}V^2 S'}.$$

Denoting the cross section of the jet $\pi d^2/4$ by S_0, the first approximation of the total induced drag of an airfoil in an air jet of circular cross section becomes

$$D = \frac{L^2}{4\frac{\rho}{2}V^2}\left(\frac{1}{S'} + \frac{1}{2S_0}\right).$$

For the solid walled channel it was stated that the circulation around the inverted wing has to have the same sign as the circulation round the actual wing. This results in a decrease in the induced drag due to the channel walls which is of the same amount as the increase in drag with the free jet. Therefore the approximate formula for the total induced drag in the tunnel is

$$D = \frac{L^2}{4\frac{\rho}{2}V^2}\left(\frac{1}{S'} - \frac{1}{2S_0}\right).$$

In order to obtain an appreciation of the numerical value of this correction, we consider a wing of which the span is equal to half the jet diameter or $S_0 = 4S'$. Here the correction equals 12.5 per cent of the induced drag. In order to determine the drag in the free atmosphere, this amount has to be subtracted from test results in the jet. The more exact formula for the correction gives 0.1262 instead of 0.125. It is seen therefore that for most practical cases the approximate formula is sufficiently accurate not only for elliptical wings but also for wings with constant lift distribution where the more exact value gives 0.127.

Glauert[1] has made an analogous calculation for the influence of channels of rectangular cross section.

In case of wings of great chord dimension with respect to the diameter of the wind tunnel, the variation of the secondary velocities along the chord cannot be neglected. A theory taking account of this effect has also been developed with the result that the wing model under test in a jet has to be given a slightly increased curvature as compared with the original wing in free air.

[1] GLAUERT, H., The Interference of Wind-channel Walls on the Aerodynamic Characteristics of an Aerofoil, *Repts. and Mem. Nat. Adv. Comm. Aeronautics (London)*, vol. 1, p. 118, 1923–1924.

CHAPTER VII

EXPERIMENTAL METHODS AND APPARATUS

A. PRESSURE AND VELOCITY MEASUREMENTS

123. General Remarks on Pressure Measurement in Fluids and Gases.—When measuring the pressure at a point in the interior of a liquid or gas, it is impossible to avoid the insertion of a foreign body, namely, the measuring apparatus, into the fluid at that point. For static conditions this is not important since the state of pressure of the fluid in the direct vicinity of the measuring instrument is not disturbed by it. On account of the finite dimensions of the instrument the average pressure over a small area is measured instead of the exact pressure at a point. This value, however, can be approximated by making the apparatus sufficiently small.

The conditions are fundamentally different when the fluid is in motion because in that case the velocity and pressure are disturbed in the neighborhood of the instrument. For instance, there must be a stagnation point on the apparatus where the fluid velocity is zero so that the pressure measured at this point would be $\rho/2 \cdot w^2$ too large, w being the undisturbed velocity (see Art. 2).

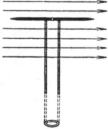

Fig. 184.—Disk for measuring static pressure.

124. Static Pressure.—If the velocity w is so large that the stagnation pressure $\rho/2 \cdot w^2$ cannot be neglected with respect to the pressure in the undisturbed fluid (the static pressure), it is not sufficient to decrease the dimensions of the instrument, but it is necessary to make its shape such that the flow is disturbed as little as possible.

A shape as shown in Fig. 184 serves the purpose. A thin circular tube is closed at the top by a very thin disk pierced in the middle. If the disk is placed in the direction of the velocity, the flow at the location of the hole is hardly influenced by it so that the pressure at this point is the same as if the disk did not

226

exist. However, if the disk is inclined under a small angle with respect to the direction of flow, a marked influence is felt and the measured pressure does not correspond to the pressure of the undisturbed fluid at that point.

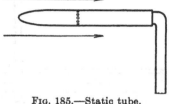

Owing to this great sensitivity against angular deviations the disk is hardly ever used any more.

The static tube shown in Fig. 185 is more advantageous in this respect. It is a thin tube held parallel to the flow with a number

FIG. 185.—Static tube.

of small holes in the side. Here the recorded pressure is much less dependent on the angular position, which will be discussed in detail in Art. 126.

A relatively simple problem is the measurement of the pressure at a solid wall along which the fluid flows, since in this case it is

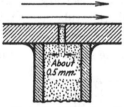

FIG. 186.—Hole in wall for measuring static pressure.

not necessary to introduce a foreign body into the flow. A small hole is drilled in the wall at the point where the static pressure is to be measured, as shown in Fig. 186. The fluid flows past the hole but remains at rest in the hole itself if its dimensions are sufficiently small. The fact that the velocity is different outside and inside the hole is not in contradiction to Bernoulli's equation, since the Bernoulli constant for the two regions is different (see Art. 58, "Fundamentals"[1]).

Owing to the influence of viscosity there is some sucking action which becomes smaller with decreasing hole diameter. According to Fuhrmann[2] the actual pressure exceeds the measured pressure by about 1 per cent of the stagnation pressure for a hole diameter of about $\frac{1}{32}$ in. In other words, if p_0 denotes the actual pressure, p the measured pressure, and w the velocity of the fluid flowing along the hole, we have

$$p_0 = p + 0.01\frac{\rho}{2}w^2.$$

[1] See footnote, p. 3.

[2] FUHRMANN, G., Theoretical and Experimental Investigations on Balloon Models (German), Dissertation, Göttingen, 1912; *Jahrb. Motorluft-schiffstudiengesellschaft*, vol. 5, p. 63, 1911–1912.

It was seen that for the measurement of static pressure it is of importance that the flow is not disturbed at the point where the pressure is to be measured. For instance, it is necessary to make sure that no burr exists at the mouth of the hole. Small inaccuracies in this respect lead to completely false results. It is therefore recommended to shape the hole approximately as shown in Fig. 186.

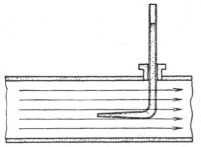

FIG. 187.—Pitot tube for measuring total pressure.

125. Total Pressure.—The total pressure, *i.e.*, the sum of the static pressure and the stagnation pressure can be measured much more easily than the static pressure by itself. Introducing into the flow an open tube, as shown in Fig. 187, causes the velocity of the fluid to become zero in the opening of the tube so that, according to Bernoulli's equation, in this stagnation point the pressure is increased by $\frac{\rho}{2}w^2$. If p_s denotes the static pressure in the stagnation point, it is seen that the instrument measures the total pressure $p_t = p_s + \frac{\rho}{2}w^2$. This tube is known as the "Pitot tube" after its inventor.[1]

It is evident that if the static pressure is known, a measurement of the total pressure immediately allows of a calculation of the velocity w, namely,

$$w = \sqrt{\frac{2}{\rho}(p_t - p_s)}.$$

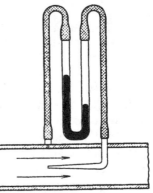

FIG. 188.—Measurement of dynamic pressure by means of mercury.

This method of velocity measurement is used very often. For instance, if the velocity at any point in the interior of a fluid or gas flowing through a pipe has to be measured, a Pitot tube is inserted into it giving the total pressure (see Figs. 188 and 189). Since the static pressure is

[1] PITOT, Description of a Machine for the Measurement of Velocity of Flowing Water (French), *Mém. acad. sci.*, p. 172, 1732.

constant across the cross section, it can be measured at the pipe wall as described before. The difference between the total pressure and the static pressure can then be measured immediately as shown in the figures.

126. Velocity Measurement with Pitot-static Tube.—In order to measure a velocity directly, an instrument has been designed in which the static pressure and the total pressure can be measured at the same time. Such an apparatus was first used by D. W. Taylor[1] and consists of a combination of a Pitot tube with a static tube. For an understanding of the limitations of this instrument the pressure distribution around a blunt-nosed hollow cylinder

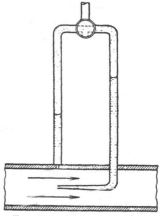

Fig. 189.—Measurement of dynamic pressure by means of the fluid itself.

is of importance. This distribution can be found experimentally (see Art. 85) by drilling into the hollow cylinder a number of very small holes which are all sealed up with the exception of one of them. The interior of the cylinder is then connected to a manometer. Inserting the cylinder into a flowing fluid with velocity w and measuring the pressures on the various holes one after another, the pressure distribution as depicted in Fig. 64 is found. Therefore, if the pressures are measured at the stagnation point of the cylinder as well as at the point where there is static pressure in Fig. 63, and if these two pressures are then connected to a manometer, the difference

$$p_t - p_s = \frac{\rho}{2}w^2$$

will be found at once. For practical reasons, however, the apparatus is made in a somewhat different form. Figure 64 or 190 shows that a very small error in the location of the static holes leads to a considerable error in the pressure. Therefore it is more accurate to locate these holes farther away from the nose of the cylinder where the vacuum decreases asymptotically toward zero. Considering further that the stem of the apparatus causes an increased pressure distributed in the manner shown in

[1] TAYLOR, D. W., *Heating Ventilating Mag.*, p. 21, 1905.

Fig. 190, it is comparatively easy to find a location where the
vacuum due to the nose of the cylinder is equal to the increased
pressure due to the stem.

It has thus been found that the shape and the dimensions
of the component parts of the apparatus are of importance for the

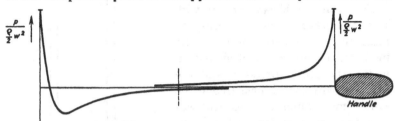

FIG. 190.—Pressure distribution on a blunt body, considering also the effect of
the stem.

results to be obtained.[1]   The dimensions given in Fig. 191, which
are due to Prandtl, have given good results.

It is to be noted that, for a flow in which the velocity oscillates
rapidly about a certain mean value in magnitude but has a

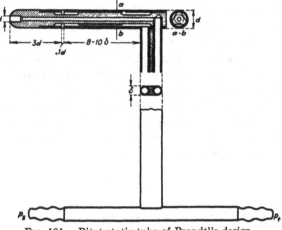

FIG. 191.—Pitot-static tube of Prandtl's design.

constant direction, the reading of the manometer or its mean
value does not correspond to the mean value of the velocity, since
the apparatus measures pressures which are proportional to the
squares of the velocity.   This point may become important in
the measurement of velocities in turbulent flows (see Art. 33).

<hr>

[1] KUMBRUCH, H., Measurement of Flowing Air by Means of Pitot-static
Tubes (German), *Forschungsarbeiten V. D. I.*, vol. 240, 1921.

The Prandtl tube is a very reliable instrument since its readings are little dependent on the angle α with respect to the direction of flow. As is shown in Fig. 192, the total pressure p_t as well as the static pressure p_s varies considerably with a change in angle α but in such a manner that their difference $p_t - p_s$, which determines the velocity, is hardly affected for angles up to $\alpha = 17$ deg.

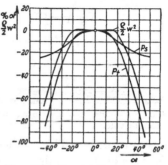

FIG. 192.—Sensitivity to changes in direction of Prandtl's tube.

Another form of Pitot-static tube, which is extensively used in the United States and England, is the one due to Brabbée (Fig. 193). This apparatus, as well as the one of Prandtl, has a proportionality factor 1 so that it does not need any calibration. It is slightly more sensitive to angular deviations than Prandtl's instrument.

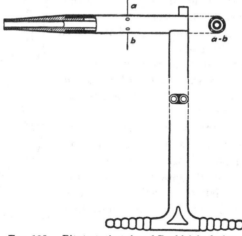

FIG. 193.—Pitot-static tube of Brabbée's design.

The influence of the turbulence of the flow on the readings of the instrument will not be discussed in detail here, but the reader is referred to the publication by Kumbruch.[1] It appears that for any form of instrument the reading is about 4 per cent high for very turbulent flows.

[1] See footnote, p. 230.

127. Determination of the Direction of the Velocity.—The measurement of the direction of the velocity is relatively

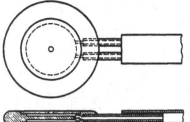

complicated. The apparatus shown in Fig. 194 consisting of a disk with openings on either side shows a difference between the pressures on the two faces except when the direction of the disk coincides with the direction of the flow. Therefore if the disk is held along the direction of the velocity, a pressure gauge shows

Fig. 194.—Disk for measuring direction of velocity and static pressure.

a zero reading. This method, however, is of use only for velocity fields which are fairly constant over large regions. In cases where the velocity varies considerably from point to point, the instrument of Fig. 195 has to be used, consisting of two Pitot tubes under 90 deg. If the manometer does not show any deviation, the direction of the flow is under 45 deg. with either tube. The relation between the manometer reading and the direction of the flow has to be found by calibration.[1]

Other instruments for the measurement of velocity will be considered later. First, some methods of pressure measurement will be discussed.

128. Fluid Manometers.—If the two water pressures to be measured are connected by means of rubber tubing to the legs of a U-shaped glass tube containing mercury, of which the specific gravity in water is γ_M,[2] the equilibrium condition is (Fig. 196)

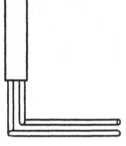

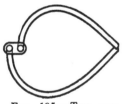

Fig. 195.—Two perpendicular Pitot tubes for determining direction of velocity.

$$h = \frac{p_t - p_s}{\gamma_M}.$$

[1] LAVENDER, T., A Direction and Velocity Meter for Use in Wind-tunnel Work, *Repts. and Mem. Nat. Adv. Comm. Aeronautics (London)*, No. 844, 1923.

[2] The term "specific gravity in water" in this connection means the specific gravity minus the buoyancy due to water, *i.e.*, $\gamma_M = \gamma_{\text{mercury}} - \gamma_{\text{Water}}$.

If γ_W denotes the specific gravity of the flowing fluid (water), we have

$$p_t - p_s = \frac{\rho}{2}w^2 = \frac{\gamma_W}{2g}w^2,$$

so that

$$h = \frac{\gamma_W}{\gamma_M}\frac{w^2}{2g},$$

and, since for water $\gamma_W = 1$,

$$w = \sqrt{2gh\gamma_M}.$$

Considering that the specific gravity of mercury is 13.6 and consequently that $\gamma_M = 12.6$, we have for $g = 386$ in./sec^2:

$$w = \sqrt{2 \times 386 \times 12.6h} = 98.6\sqrt{h} \text{ in./sec}$$
(velocity of water with mercury as manometer fluid).

Therefore a level difference in the manometer of 4 in mercury corresponds to a velocity of about 16 ft/sec. Assuming that a level difference of 0.01 in. is the limit of accuracy of the manometer, the smallest velocity that can be measured with this method is about 10 in./sec. Using the water itself as the manometer fluid (Fig. 189), we have

$$w = \sqrt{2 \times 386 \times h} = 27.8\sqrt{h} \text{ in./sec}$$
(velocity of water with water as
manometer fluid).

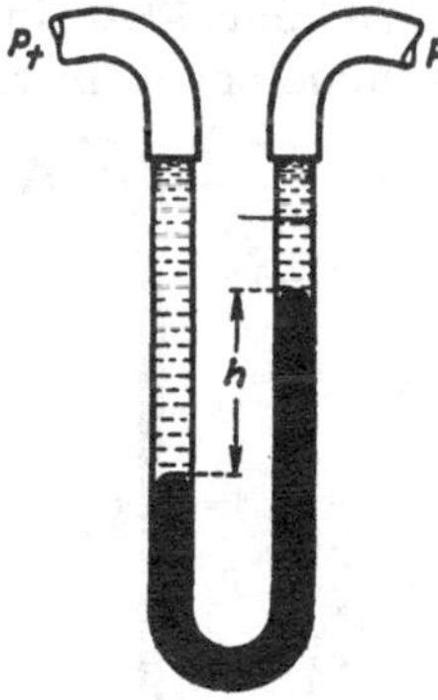

Fig. 196. — V-tube manometer.

For pressure measurements or velocity measurements in gases, the method remains practically the same; only a manometer fluid of small specific gravity, for instance water or alcohol, is used for the usual gas pressures. Since the specific gravity of water referred to air of room temperature and usual barometric pressure is

$$\frac{\gamma_{\text{Water}}}{\gamma_{\text{Air}}} = 800,$$

we have

$$w = \sqrt{2 \times 386 \times 800 \times h} = 780\sqrt{h} \text{ in./sec}$$
(velocity of air with water as manometer fluid).

A level difference of 0.01 in. water therefore corresponds to an air velocity of 6.5 ft/sec., which shows that for small air velocities the level difference has to be measured very accurately.

With the usual U-tube mercury manometer without special optical appendages, level differences can be estimated down to about 0.004 in., whereas for water on account of capillary phenomena the reading cannot be trusted with any better accuracy than about 0.04 in. Therefore in all cases where accuracy is required, the water of the manometer is replaced by organic fat-dissolving fluids, like alcohol, toluol, etc. If γ is the specific gravity of this manometer fluid as compared to water, the above formula for velocity has to be changed to

$$w = 780\sqrt{h\gamma}.$$

With these organic fluids, the accuracy is increased to about 0.010 in. so that with this method air velocities down to about 6 ft/sec can be measured. For still smaller velocities or pressures special sensitive manometers have to be used.

129. Sensitive Pressure Gauges.—The sensitivity of fluid manometers is increased either by special optical devices for observing the meniscus or by inclining the legs of the gauge at an angle. A third method which is used rather seldom consists of replacing the air column by a fluid lighter than water and not miscible with it, as for instance kerosene or amyl acetate. The instrument has to be completely filled with liquid. In this manner only the difference in the specific gravities of the two fluids is acting; for water and kerosene this is about 0.2. The method which is five times as sensitive (and could even be made more sensitive by a suitable choice of liquids) has the disadvantage that the meniscus between water, kerosene, and glass is not so distinct as between air, alcohol, and glass.

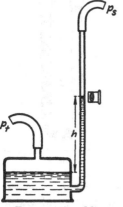

Fig. 197.—Micromanometer where sensitivity is obtained by lens system for reading height of meniscus.

With the usual sensitive manometers one of the legs of the U-tube is transformed into a basin of relatively large cross section (Figs. 197 and 198). This has the advantage that only one reading has to be made, since the change in level in the basin can be either neglected or subsequently corrected. This decreases the error of the procedure by 50 per cent. For instance, if the diameter of the basin is 4 in. and that of the other leg ½ in., the fluid level in the basin sinks $(\frac{1}{2}/4)^2$, *i.e.*, one sixty-fourth part of the change in level in the other leg. The readings of the manometer in this example therefore have to be increased by one sixty-fourth to correct for the change in level in the basin.

Modern constructions of inclined manometers are usually such that the inclined leg can be swiveled round its connection with the basin. By this device several ranges of the instrument and several regions of accuracy can be obtained. Let Δp be the pressure difference in inches of water, α the angle of the inclined leg

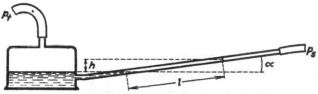

FIG. 198.—Inclined-tube micromanometer.

with respect to the horizontal, l the travel of the meniscus in inches, and γ the specific gravity of the manometer fluid, then we have (see Fig. 198)

$$\Delta p = l\gamma \sin \alpha,$$

if the level change in the basin is neglected. It is thus seen that the sensitivity becomes greater for smaller angles α. For inclinations down to about $\sin \alpha = \frac{1}{25}$, this kind of manometer can be used without any special precautions; for still smaller inclinations, however, the errors due to capillary action become more and more serious. These errors can be avoided only by very careful calibration.

Assuming that a change in the meniscus can be observed with an accuracy of 0.01 in., this travel of the meniscus with an inclination of one twenty-fifth and alcohol as a manometer fluid (specific gravity 0.8) corresponds to a pressure

$$\Delta p = 0.01 \times \frac{1}{25} \times 0.8 = 0.32 \cdot 10^{-3} \text{ in. (water).}$$

With this instrument therefore an air velocity as small as 15 in./sec can be measured.

The calibration of inclined-leg pressure gauges is accomplished by putting a carefully weighed quantity Q of the manometer fluid into the basin and observing the travel of the meniscus due to this. If A be the cross section of the basin, the amount Q increases the level by $Q/\gamma A$ and therefore is equivalent to an air pressure of $\Delta p = h\gamma = Q/A$. By repeating this procedure a number of times a complete calibration curve $p = f(l)$ can be obtained.

For very small inclinations (under one twenty-fifth) this calibration has to be done in small steps for the individual parts of the entire capillary tube in order to determine the errors of the capillary tube itself.

An improvement in this instrument giving still greater accuracy is due to Rosenmüller.[1]   His apparatus is shown schematically

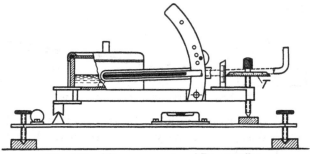

FIG. 199.—Micromanometer of Rosenmüller.

in Fig. 199.   Instead of determining the pressure difference from the travel of the meniscus (which includes all the errors of the capillary), the inclined capillary tube is swiveled until the original zero reading is established. The angle through which the capillary is turned can be read off a micrometer screw T. By suitable construction of the pitch of the screw, one division of T corresponds to 0.001-mm water pressure, which constitutes the sensitivity of the instrument. The advantages of this construction are that the irregularities of the capillary tube do not enter into the result and that the reading can be accomplished in a relatively short time (about 3 min).

If such a great sensitivity is not necessary and if a greater range of velocities is to be measured (up to about 12 in. water corresponding to about 230 ft/sec wind velocity), the precision manometer with vertical leg developed by the Aerodynamic

[1] ROSENMÜLLER, M., New Measuring Apparatus of Air Velocities (German), *Messtechnik*, vol. 2, p. 343, 1926.

Institute of Göttingen is to be recommended.[1] The sensitivity of this instrument is due to the fact that the meniscus can be observed very accurately.   Parallel to the manometer tube there is a scaled guide carrying a vernier, a lens in front of the tube, and a concave mirror M behind it (Fig. 200).   This mirror gives an inverted real image of the meniscus.   The carriage is adjusted to such a position that the actual meniscus seen through the lens

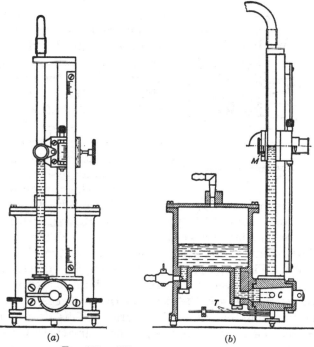

(a)   (b)

Fig. 200.—Micromanometer of Prandtl.

is just touching the inverted meniscus of the mirror, which can be done very accurately.   In this position the vernier is read by means of a second lens which allows a determination of the position of the carriage to 0.002 in. close.   For very rapidly varying pressures two different degrees of damping can be inserted by means of two capillary tubes T.   Another precision manometer with a range of 4- to 6-in. water pressure with a sensitivity of 0.0004 in. water has been brought on the market by the Askania Works, Berlin.

[1] PRANDTL, L., *Göttinger Ergebnisse*, vol. I, p. 44, Munich, 1921.

For still greater sensitivities (to about 0.04 mil water) with a smaller range (about 2 in.) special manometers have been constructed. Among these the most important one is the Chattock gauge developed in England.[1] The instrument, shown schematically in Fig. 201, consists of a glass U-tube of somewhat

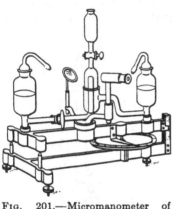

extraordinary shape which is attached to a metal frame that can be tilted round an axis. The two pressures are connected to the two reservoirs right and left, which are half filled with water (or with a salt solution of specific gravity 1.07). If the vessel in the middle would be filled with the same liquid, a flow of the salt solution from the outer vessel of greater pressure to that of lower pressure could not be detected in the middle vessel. In order to show a displacement from the high-pressure side to

Fig.   201.—Micromanometer   of Chattock.

the low-pressure side, the middle vessel is filled with castor oil, which does not mix with water.   The glass tube connecting the left vessel to the middle reservoir protrudes into the castor oil, and the salt solution forms a very distinct meniscus with the oil on the top of the tube.   This meniscus is viewed through a microscope with crossed wires.   If a very small pressure difference occurs between the two extreme vessels, the shape of the meniscus between the salt solution and the castor oil deforms.   This deformation is made to disappear by giving the proper inclination to the frame carrying the glass vessels, which is done by turning a micrometer screw from the reading of which the pressure difference can be calculated.   The sensitivity of this micromanometer is $6 \cdot 10^{-6}$ in. of water according to Chattock.

A similar manometer of the same sensitivity but with a range of 6 in. has been described by Douglas;[2] see also the paper by

[1] CHATTOCK, A. P., Note on a Sensitive Pressure Gauge being an appendix to: On the Specific Velocities of Ions in the Discharge from Points, *Phil. Mag.*, 1901, p. 79; see also J. R. Pannell, Experiments with a Tilting Manometer for Small Pressure Differences, *Engineering*, vol. 96, p. 343, 1913.

[2] DOUGLAS, G. P., Note on a Large-range Manometer for Wind-tunnel Work, *Repts. and Mem. Nat. Adv. Comm. Aeronautics (London)*, vol. 1, p. 110, 1919–1920.

Duncan.[1] A manometer of very great sensitivity (about $4 \cdot 10^{-7}$-in. water) has been described by Fry.[2]

Another manometer of rather great sensitivity has been developed in the Aerodynamic Institute, Aachen.[3] Two vessels V_1 and V_2 of accurate cylindrical shape to which the pressures are connected contain two floats which are rigidly attached to each other and carry a mirror between them. The pressure difference causes a difference in the water level between the vessels V_1 and V_2 which turns the connection between the floats and consequently the mirror. This angular deviation of the mirror is observed with a telescope. The sensitivity of the instrument is about 10^{-4} in. of water. A disadvantage of this manometer is that relatively large amounts of water have to be moved by very small forces so that it requires from 30 to 45 min to obtain one reading.

Finally, we mention an air micromanometer made by Edelmann & Sohn, Munich.[4] In this instrument, the air from the spot where its pressure is to be measured is blown through a nozzle against a small mica vane attached to a torsion wire. The angle of torsion of the mica vane is measured by means of a mirror attached to the same wire. The sensitivity is said to be about $4 \cdot 10^{-6}$ in. of water.

Recording manometers have been constructed on the principle of either the aneroid barometer or utilizing floats. Wieselsberger[5] has constructed an aneroid barometer which has also been adapted[6] to the registration of the velocity of an airplane relative to the surrounding air.

130. Vane-wheel Instruments.—Besides the Pitot tubes discussed in Art. 126, there exist a number of instruments which require calibration before they can be used.

The most important among these utilize wheels with vanes or buckets. For water measurements the usual rotary-disk water

[1] Duncan, W. J., On a Modification of the Chattock Gauge, Designed to Eliminate the Change of the Zero with Temperature, *Tech. Rept.* 1069, *Aero. Research Comm.*, 1927, p. 848, London, 1928.

[2] Fry, J. D., A New Micromanometer, *Phil. Mag.*, vol. 25, p. 494, 1913.

[3] Ermish, H., Flow and Pressure Distribution of Obstacles as a Function of Reynolds' Number (German), *Abhandl. Aero. Inst., Tech. Hochschule Aachen*, vol. 6, p. 21, Berlin, 1927.

[4] *Z. Ohrenheilk.*, vol. 56, p. 344.

[5] Wieselsberger, C., *Göttinger Ergebnisse*, vol. 2, p. 6, Munich, 1923.

[6] Wieselsberger, C., A Manometer for Recording Flying Speeds (German), *Z. Flugtech. Motorluftschiffahrt*, vol. 12, p. 1, 1921.

meters fall under this class, whereas for air-speed measurements the instruments are known as anemometers, among which we have to distinguish between vane- or wind-mill type anemometers and hemispherical-cup anemometers. The number of revolutions of the instrument is read off on a revolution counter, but there are also constructions where the wheel operates an electric bell after a certain number of revolutions. The time elapsed between two strokes on the bell is determined by means of a stop watch. The calibration of water meters of this type can be accomplished by towing them with a constant velocity through water at rest. The calibration of anemometers for small wind velocities up to 30 ft/sec is done mostly on the rotating arm (see Art. 140). The relative air velocity of the anemometer is equal to the arm velocity corrected by the wind which is caused by the moving arm. For large velocities (above 15 ft/sec) anemometers are usually calibrated in the artificial air stream of a wind tunnel and compared with the readings of a Pitot tube.

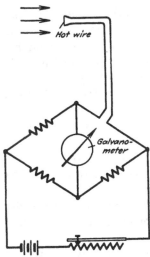

Fig. 202.—Constant-voltage hot-wire anemometer; the bridge voltage is kept constant.

Owing to the considerable inertia of the vanes, all instruments of this type indicate only the mean value of the velocity with respect to time. In wind of varying intensity the readings of the anemometer show a considerable phase lag with respect to the wind velocity.[1] Gusts of wind of very short duration cannot be measured with this kind of apparatus. In case an anemometer is used for the determination of the velocities in a pipe, it is to be considered that, owing to the volume which the instrument takes up in the pipe, the indicated velocities are higher than those in the undisturbed pipe. With the usual anemometer placed in a pipe of about 10 in. diameter this error amounts to about 3 per cent.

[1] SCHRENK, O., On the Errors Due to Inertia in the Readings of Hemispherical Anemometers in Wind of Varying Intensity (German), *Z. tech. Physik*, vol. 10, p. 57, 1929.

131. Electrical Methods of Velocity Measurement.—Another method of measuring velocities is based on the fact that an electrically heated wire exposed to the air stream cools off and consequently changes its electric resistance. This method is especially valuable for moderately small air velocities. The hot wire which is usually very thin (0.5 to 5 mils diameter) is connected in a Wheatstone bridge circuit (Fig. 202), which makes the measurement of a small change in the resistance extremely accurate. This sort of instrument is usually calibrated on the rotating arm.

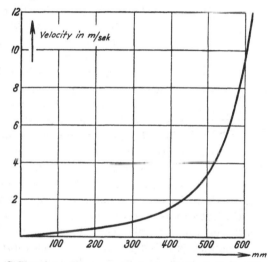

Fɪɢ. 203.—Calibration curve of a constant-voltage hot-wire anemometer.

Hot-wire anemometers are used in two kinds of circuits: (1) constant-voltage and (2) constant-resistance circuits. With the first method the voltage across the bridge is kept constant after having been adjusted to such a value that the galvanometer shows zero current when the hot wire is in still air. As soon as the air starts flowing, the hot wire cools off and the galvanometer shows a reading which is related to the wind velocity in a manner determinable by calibration. This circuit was first suggested by Weber[1] and was developed further by King.[2] It is useful only for very small air velocities, but in this range it

[1] Wᴇʙᴇʀ, L., *Schriften naturwiss. Ver. Schleswig-Holstein*, vol. 2, p. 313, 1894.

[2] Kɪɴɢ, R. O., *Phil. Trans. Roy. Soc. (London)*, A, vol. 214, p. 373, 1914; *Phil. Mag.*, vol. 29, p. 556, 1915; *J. Franklin Inst.*, vol. 181, p. 1, 1916.

is extremely sensitive, the velocity being determinable down to 0.2 in./sec.[1] Since the very thin platinum wire, which

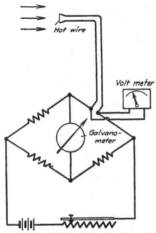

is usually heated to a dull-red heat, is cooled off considerably by rather small air velocities, a further increase in the velocity results only in a relatively slight further cooling and consequent change in resistance. Owing to this fact, the method of constant voltage is not very sensitive for larger air velocities. Figure 203 shows the relation between air velocity and galvanometer reading for an instrument with 4-mils wire diameter.

Fig. 204.—Constant-resistance hot-wire anemometer; the resistance and consequently the temperature of the hot wire are kept constant by varying the bridge voltage.

With the second method the resistance in the battery across the bridge is increased to such a value that the wire which was originally cooled off by the air current is again brought up to its first temperature (Fig. 204).

The temperature of the hot wire and consequently its resistance are kept constant by varying the bridge voltage so that the galvanometer reading remains zero. The current in the hot wire is read by means of a voltmeter which gives a measure for the air velocity. This method has been improved by Callendar (see King[2]) so that the calibration curve of the instrument is practically a straight line even for very small air velocities. The calibration curves for the various types of hot-wire anemometers are shown in Fig. 205.

Another method for obtaining a practically straight-line characteristic is by using a compensating hot wire, as shown in Fig. 206.[3] The compensating hot wire H_1H_2 is always kept in still air and the resistor is adjusted to such a value that if the main hot wire is also in still air the galvanometer reading is zero. When the air begins to flow, the galvanometer gives a certain reading which is practically proportional to the wind velocity, as

[1] OVERBECK, A., *Ann. Physik*, vol. 56, 397, 1895; DAU, R., Dissertation, Kiel, 1912.

[2] KING, R. O., The Measurement of Air Flow, *Engineering*, vol. 117, pp. 136, 249, 1924.

[3] HUGUNEARD, MOGNAU, and PLANIOL, On a Compensated Hot-wire Anemometer (French), *Compt. rend.*, vol. 176, p. 287, 1923.

shown by curve *a*, Fig. 207; curve *b*, for a constant-voltage hot-wire instrument, is shown for comparison.

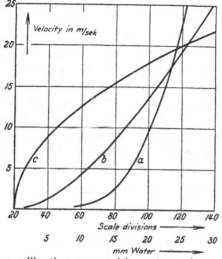

Fig. 205.—Various calibration curves: (*a*) constant-voltage anemometer; (*b*) constant-resistance anemometer; (*c*) dynamic-pressure curve.

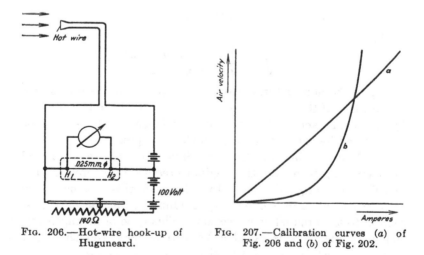

Fig. 206.—Hot-wire hook-up of Huguneard.

Fig. 207.—Calibration curves (*a*) of Fig. 206 and (*b*) of Fig. 202.

For the special purpose of investigating the structure of the wind the Siemens & Halske Company in Berlin has put on the market an ingenious hot-wire recording instrument designed by Gerdien.[1]   This instrument automatically records the wind velocity with its smallest and fastest variations as well as its horizontal direction and its vertical component.

132. Velocity Measurements in Pipes and Channels.—In case the mean velocity of gas or water in pipe lines has to be determined (with a view toward finding the transported volume), it is possible to determine the velocity in a good many points of the cross section by means of a Pitot tube or a hot-wire anemometer. This method is very laborious, especially for non-circular cross sections, and, moreover, its accuracy is not great on account of the rapid drop in velocity near the wall of the pipe.

The method of finding the mean velocity by means of the pressure variations due to cross-sectional variations has been found more practical. Bernoulli's equation states that the pressure p_0 is decreased by an amount $\frac{\rho}{2}(w^2 - w_0{}^2)$ when the velocity is increased from w_0 to w:

$$p_0 - p = \frac{\rho}{2}(w^2 - w_0{}^2).$$

If the cross-sectional area drops from A to a, the continuity equation is

$$w_0 = \frac{a}{A}w,$$

so that

$$w = \sqrt{\frac{1}{1 - a^2/A^2}} \sqrt{\frac{2(p_0 - p)}{\rho}}.$$

Owing to the non-uniform velocity distribution in the pipe above the location of the measurement, this velocity has to be corrected by a certain factor, the "velocity coefficient," which is to be determined by calibration for each shape of pipe.

133. Venturi Meter.—Certain difficulties are encountered in attempting to restore the original pressure by decreasing the velocity to its original value.   In order to do this, it is necessary to increase the cross section very gradually from the narrowest section to the original cross section.   This type of arrangement, shown in Fig. 208, is called a Venturi meter.   Herschel[2] first

[1] GERDIEN, H., The Anemoklinograph, an Apparatus for the Investigation of the Structure of the Wind (German), *Jahrb. wiss. Gesellsch. Flugtech.*, vol. 2, 1913–1914.

[2] HERSCHEL, CL., The Venturi Meter, paper read before the *Am. Soc. Civil Eng.*, December, 1887.

suggested its use for the measurement of delivered volume in pipe lines. In order to find the relation between the pressure difference and the mean velocity in the pipe a calibration curve of a geometrically similar Venturi meter has to be known, and in cases where the velocity of approach is not very small with respect to the velocity in the throat this geometrical similarity has to be extended to the approach as well. For Venturi tubes of the shape shown in

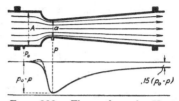

FIG. 208.—Flow through Venturi tube; the full line gives the pressure distribution along the center line; the dashed curve along the wall.

Fig. 208 the velocity coefficient is approximately 1.00.

134. Orifices.—In spite of the fact that with a Venturi meter the pressure drop is very small (about 15 to 20 per cent of the

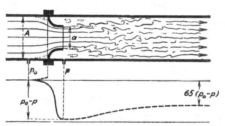

FIG. 209.—Flow through rounded-approach orifice, full and dashed lines as in Fig. 208.

pressure drop in the throat), its practical application is limited by its large size. Therefore standardized orifices as shown in Figs. 209 and 210 are used more frequently. The pressure

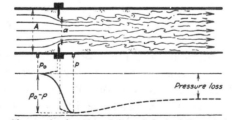

FIG. 210.—Sharp-edged orifice; full and dashed lines as in Fig. 208.

diagrams in these two figures show that with this kind of apparatus the loss in pressure is from 60 to 70 per cent of the pressure drop in the orifice.   The velocity coefficient α has been found to be 0.96 to 0.98 with the standardized (German) rounded-approach orifice (Fig. 209).   For the sharp-edged orifice shown in Fig. 210 the coefficient depends very much upon the ratio of the cross sections a/A.   For instance, for $a/A = 0.15$, we have $\alpha = 0.61$, whereas for $a/A = 0.75$, the velocity coefficient is $\alpha = 0.91$.[1]

135. Weirs.—For the measurements of velocity in open channels, weirs are used most frequently.   The height h of the undis-

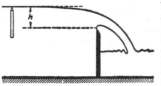

FIG. 211.—Flow over weir.

turbed water level above the crest of the weir is a measure for the discharge per unit of time (Fig. 211).   It is necessary to ventilate the weir, *i.e.*, to let air pass freely under the jet.   In the absence of this precaution a vacuum will be created under the jet, which will pull the jet down and increase the discharge.   With ventilated weirs an accuracy of 99.5 per cent can be obtained.[2]

Since the difference in height between the water level and the crest of the weir in general is small, its measurement has to be carried out with precision.   The usual method is to have a micrometer screw with a sharp conical point entirely submerged in the water.   This point is screwed up until it touches the water surface, which can be observed very accurately since at that moment the point itself and its reflected image coincide on the surface.   The observation is made from below through a glass window in the side of the tank.

136. Other Methods for Volume Measurement.—For small volumes of either water or gas, the method of direct weighing is useful.   The amount discharged from the pipe is collected in a suitable vessel during a definite interval of time and then either the volume or the weight is accurately determined.   For gases the possible error due to temperature changes has to be considered.

Besides the method of direct weighing, volumes in small quantities can be measured by ordinary domestic water or gas meters as well as by the method of salt titration.

[1] Volume Measurement with Standardized Orifices (German), V. D. I., Berlin, 1930; MUELLER, H., and H. PETERS, Correction Factors for Standardized Orifices (German), Z. V. D. I., vol. 73, 1929.

[2] REHBOCK, TH., Discharge Measurements with Sharp-crested Weirs (German), Z. V. D. I., vol. 73, p. 817, 1929; DE THIERRY, G., and C. MATSCHOSS, "The Hydraulic Laboratories of Europe" (German), p. 104, Berlin, 1926.

B. DRAG MEASUREMENTS

137. The Various Methods.—Because of the fact that the drag on a body is the same whether the body is at rest and the air is moving or whether the body is moving and the air is at rest, there are two different methods of drag measurement.

The first method, where the fluid is at rest, again has several modifications.[1]

The body may be attached to a carriage running on rails and towed through the fluid, or it may be permitted to fall down freely, guided only by a vertical guide wire, or again it may be mounted on the extremity of an arm which is rotated through still air.

138. Towing Tests.—The method of towing the test specimen is restricted practically to water, as experience has shown it to be impractical for air. This is due to the fact that the carriage on which the body is mounted is also moving through the air and generally creates considerable disturbance in it. In case the experiment is conducted in the open, the irregularities of the free outside atmosphere are also very disturbing. Moreover, it is difficult to move the carriage with an accurately constant velocity. Any deviation from constant velocity will cause inertia forces in the test specimens, which may become of the same order of magnitude as the wind reactions. For a practical realization of this method the test track has to be very long, which makes the construction as well as the operation of such apparatus expensive.

However, for drag measurements in water the method has been used very successfully, primarily of course in connection with the problem of ship resistance. Experimental tanks for this purpose can be found in many laboratories all over the world.

139. The Method of Free Falling.—This procedure has been worked out only for air. The first experiments in this direction we owe to Piobert, Morin, and Didion (1835),[2] who reached velocities up to 30 ft/sec. They had an apparatus for recording the velocity. As soon as the velocity had become uniform, the drag was equal to the weight. The method was greatly improved by Cailletet and Colardeau (1892).[3] They dropped plane sur-

---

[1] The various methods used prior to 1910 are described in detail by G. Eiffel, "The Resistance of the Air" (French), Paris, 1910.

[2] Memoirs on the Laws of Air Resistance (French), *Mémorial de l'Artillerie*, No. 5, 1842.

[3] *Compt. rend.*, vol. 115, p. 13, 1892.

faces of various shapes down from the Eiffel Tower and measured the relation between the time required and the height of fall. They reached velocities up to 90 ft/sec. Eiffel[1] brought this method to complete development (1905) and made an elaborate series of tests on various bodies with velocities up to 130 ft/sec.

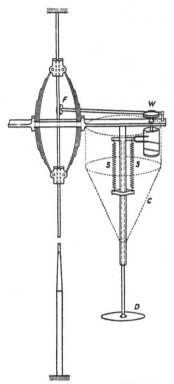

His apparatus is shown schematically in Fig. 212. The whole apparatus is sliding down freely on two bearings along a tightly stretched vertical wire. The object under test, D, is attached to two springs SS, which expand proportionally to the drag of D. The extension of the springs is recorded on a rotating drum by means of a tuning fork. The drive of this drum is by means of a worm W and a friction wheel F. Owing to the vibrations of the tuning fork, the record is not a smooth curve but has little ripples on it which indicate the time. The abscissa of the drum record is proportional to the height of the fall and the ordinate is proportional to the drag. At the end of the fall the guiding wire becomes thicker so that the apparatus is brought to a stop. The entire mechanism is balanced by means of a body of small resistance C, which is drawn in the sketch in dotted lines but really is situated on the other side of the wire. If Q is the weight of the test specimen and of all moving parts attached to it (tuning fork), w the velocity, and f the spring force, the drag D is

Fig. 212.—Apparatus of Eiffel for falling experiments.

$$D = f + Q\left(1 - \frac{1}{g}\frac{dw}{dt}\right),$$

since the product of mass and acceleration of the specimen D must be equal to the sum of all forces acting on it. The values

---

[1] EIFFEL, G., "Experimental Researches on Air Resistance Conducted at the Eiffel Tower" (French), Paris, 1907.

of f and of dw/dt can be read from the diagram. The disadvantage of all fall methods in general is that the recording apparatus has to move with the specimen. This is bound to affect the flow conditions behind the obstacle, which, as we know, are of great importance.

140. Rotating-arm Measurements.—The method of drag measurement by the revolving arm has been used by various investigators, especially during the last century. The pioneer of aeronautics, O. Lilienthal,[1] made his fundamental experiments (1870) in this manner on flat and curved plates. Figure 213 shows the apparatus used by him, which was also capable of

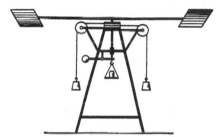

Fig. 213.—Rotating-arm apparatus of Lilienthal.

measuring the lift. The drive by means of falling weights is primitive in comparison with the later constructions of Langley[2] and Dines;[3] this was due, however, to the fact that Lilienthal had hardly any money for conducting his experiments, on which account he deserves all the more credit for his fundamental researches.

The main disadvantage of the rotating-arm method is that after one-half revolution the plate or obstacle does not pass any more through still air but through the wake of the other plate, which generally consists of very turbulent air. Moreover, owing to the rotation of the arms, the air is gradually put into rotation itself. This additional air velocity was recognized by Lilienthal but not considered in his calculations. It would be necessary

[1] LILIENTHAL, O., "The Flight of Birds as the Foundation of the Art of Flying" (German), Berlin, 1889.

[2] LANGLEY, S. P., The Internal Work of the Wind, *Phil. Mag.*, vol. 37, p. 425, 1897.

[3] DINES, W. H., Some Experiments Made to Investigate the Connection between the Pressure and Velocity of the Wind, *Quart. J. Meteorolog. Soc.*, vol. 15, 1889.

to measure this relative velocity by means of very sensitive vane-wheel or hot-wire anemometers. Another difficulty of this method is to take care of the action of centrifugal acceleration on the flow. Because of all these factors combined, the drag measurements with the rotating arm show serious errors and therefore the method is now hardly ever used.

Another method which has become obsolete utilizes a pendulum and is associated with the names of Borda, Hergesell, and Frank. In this case the motion naturally is accelerated or decelerated all the time, and the velocities involved are very small. It has the same disadvantages as the rotating-arm method, namely, that the test specimen is moving through the turbulent air of its own wake.

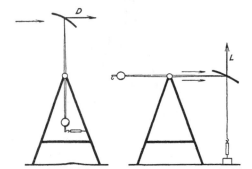

Fig. 214.—Lift- and drag-measuring apparatus of Lilienthal.

Now we shall proceed to discuss the second general method of drag measurement, where the test specimen is at rest and the air is streaming with respect to it.

141. Drag Measurement in the Natural Wind.—At first thought the simplest way of measuring the drag of a test specimen seems to be to subject it to the action of the natural wind, the velocity of which can be measured by means of one of the methods discussed before. In fact, this is the oldest procedure known. Besides his measurements on the rotating-arm apparatus, O. Lilienthal has made lift and drag determinations on planes inclined slightly with respect to the natural wind. The apparatus used by him is shown schematically in Fig. 214. He found that the drag values obtained in this manner were considerably different from those obtained by means of the rotating arm. Several commentators on Lilienthal's work concluded from this discrepancy that it made a difference whether the test specimen was moved with respect to the surrounding air or whether the air was moved with respect to the object. However, the matter can be easily explained by the fact that both methods are inherently very inaccurate.

The drag measurement in a natural wind differs in two fundamental points from the method of towing or of free falling. The free wind is always more or less turbulent, whereas an object towed through still air experiences a laminar flow at least at its front side. The other fundamental difference lies in the fact that the wind intensity depends very much on the distance from the ground. Exactly at the surface of the ground the wind velocity is zero, from which value it increases rapidly with the height above the ground. Very small hills or other unevennesses of the ground are capable of upsetting the test results completely. The natural wind, moreover, is seldom very steady in its magnitude but is always more or less gusty.

142. Advantages of Drag Measurement in an Artificial Air Stream.—In order to avoid the various difficulties of the free atmospheric wind, it has become customary to create artificial air streams by means of blowers and to study their action on the models. The advantages of this method are obvious. In the first place the various components of the force acting on the test model can be measured by means of sensitive scales, one after the other. For these measurements plenty of time can be taken, which improves their accuracy, whereas in the free wind all results have to be taken from recording instruments which are inherently less accurate. An advantage over the falling method is that all difficulties relating to inertia forces due to the necessary acceleration period are avoided. Furthermore it is possible to locate all measuring instruments outside the air stream, while the test model itself is held in place in the stream by means of thin wires or struts. For this reason, the errors arising from the fact that the air stream is affected not only by the test specimen but also by its supports are reduced to a minimum.

With any measurement employing artificial air streams, it is of great importance, however, that the stream reach the model in as uniform a state as possible. We shall now proceed to a discussion of the various test set-ups in existence and shall examine to what extent the requirement of non-turbulence of the stream is satisfied.

C. WIND TUNNELS

143. The First Open Wind Tunnels of Stanton and Riabouchinsky.—The first wind tunnel was built by Stanton[1] in the

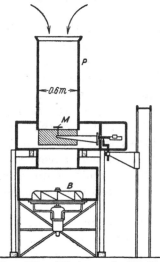

National Physical Laboratory in London in 1903 (Fig. 215). The air was sucked by a ventilator B through an intake tube P and then flowed past the test model M. At this point the tube widened out to a box in which a very sensitive scale was mounted. The model was attached to one arm of the scale by means of a very thin strut. The maximum air velocity was 30 ft/sec; the diameter of the air stream 2 ft. Stanton and, after him, Riabouchinsky found that even for small test specimens the drag is affected by the walls of the tube. For plates of more than 2 in. width (*i.e.*, about 8 per cent of the tube diameter), the drag per unit area increased rapidly with the width of the plate.

FIG. 215.—Wind tunnel of T. E. Stanton (1903).

On suggestions from Joukowsky a very elaborate experimental laboratory was built in 1906 in Moscow by Riabouchinsky.[2] The tunnel had a diameter of 4 ft and a length of about 45 ft. The model was suspended in the middle of this tunnel, where the cylindrical walls were made of glass so that the model could be observed during the test. The vorticity of the air was diminished by arranging a rather large intake nozzle at the entrance of the tunnel and also by installing a number of honeycomb grids, with the result that the air velocity had variations less than 4 per cent of the mean. The wind velocity could be varied from 3 ft/sec to 20 ft/sec. As in Stanton's tunnel, the air was also sucked in by a blower since it had been shown by previous experiments that an air stream of this kind is far less turbulent than one blown into the tunnel.

[1] STANTON, T. E., On the Resistance of Plane Surfaces in a Uniform Current of Air, *Proc. Inst. Civil Eng.*, vol. 156, London, 1903–1904.

[2] RIABOUCHINSKY, D.: Bull. Inst. aérodynamique de Koutchino, vols. I, II, and III, Moscow, 1906, 1909.

144. The First Closed Wind Tunnels in Göttingen and London.—Prandtl in Göttingen (1907–1909) and Stanton in the National Physical Laboratory in London (1910) constructed closed wind tunnels where the air discharged by the ventilator is guided through a closed circuit and after having been freed from vorticity is led back to the test model. The Göttingen tunnel of 1909[1] was intended to be a temporary one in order to obtain experience for a subsequent larger construction. It was demolished in 1918 and reconstructed in a somewhat modified form.[2]

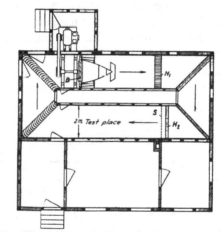

Fig. 216.—First closed wind tunnel of Prandtl (1907–1909).

The old tunnel had a cross section of 6 by 6 ft. The wind velocity could be varied up to 30 ft/sec (Fig. 216). In order to guide the air four times through a right angle, special guide vanes were built in. The apparatus for smoothing out the air after leaving the ventilator consisted of two honeycomb rectifiers H_1 and H_2, coarse and fine, respectively, and of a sieve S of 0.1-in. opening. After having passed through these, the air struck the test model, which was supported by means of thin wires on the aerodynamic balance. Near the model the wall of the tunnel had windows so that observations could be made during the test. It was found that the rectifiers and the sieve did not insure sufficient uniformity of the velocity across the section of the tunnel and,

[1] Prandtl, L., The Importance of Model Experiments for Aeronautics and the Apparatus for Such Tests in Göttingen (German), *Z. V. D. I.*, vol. 53, p. 1711, 1909.

[2] *Göttinger Ergebnisse*, vol. 2, p. 1, Munich, 1923.

in order to secure this uniformity, a process of correction, consisting of widening or narrowing some of the openings, was applied to both rectifiers. A uniformity of the air velocity within ± 1 per cent was thus obtained. When discussing the larger wind tunnel in Göttingen, Art. 147, it will be seen that there are better methods of obtaining uniformity in the air velocity with simpler means.

The wind tunnel built by Stanton[1] in 1910 is shown schematically in Fig. 217. The air is sucked through the inside channel by a blower B. It then passes through the outside channel and returns to the inside one, in which the test models are suspended. At the entrance of this test tunnel, which has a cross section of 4 by 4 ft, a honeycomb is provided in order to

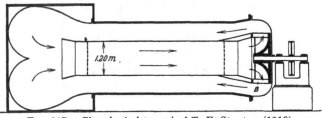

Fig. 217.—Closed wind tunnel of T. E. Stanton (1910).

smooth out the air stream. The attachment of the models to the balance is by means of a thin strut, such as was used on the previous English construction (Fig. 215).

145. The First Wind Tunnel of Eiffel with a Free Jet.—The construction of a free-jet wind tunnel, first introduced by Eiffel[2] in 1909, constitutes a definite improvement. The walls of a wind tunnel prevent the air from flowing freely around an object of somewhat large size. In order to avoid this effect, Eiffel replaced the tunnel walls for a short stretch near the test model by a large air-tight chamber (Fig. 218). This construction has the added advantage that the models can be approached without difficulty at any time.

Another advantage of the free jet over the closed channel is that the pressure along the length of the jet is practically constant, equal to the pressure of the surrounding air, which con-

[1] STANTON, T. E., Report on the Experimental Equipment of the Aeronautical Department of the National Physical Laboratory, *Rept. Adv. Comm. Aeronautics*, 1909–1910, London, 1910.

[2] EIFFEL, G., "The Resistance of Air and Aviation" (French), Paris, 1910.

sequently makes the air velocity in the jet constant (with the exception of the narrow range near the boundary, where mixing with the outside air takes place). With a channel, on the other hand, the boundary layer increases in thickness along it in the direction of the flow. This tapers the cross section for the undisturbed air stream down to a smaller diameter, which leads to an increased velocity along the stream. The air is sucked from the hall H through a nozzle N, a sieve, the test chamber T, and a receiving nozzle by means of a ventilator B. The air is then pushed through a channel of widening cross section D back into the hall H. The model is suspended in the test chamber

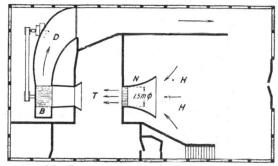

FIG. 218.—The first wind tunnel with free jet of Eiffel (1909).

about 3 ft distant from the sieve. The velocity of the air stream can be varied between 15 and 70 ft/sec approximately. Since the air in the hall H has atmospheric pressure, the pressure in the test chamber is lower; according to Bernoulli's law, this difference is about 1 in. water for 70 ft/sec wind velocity. Since the jet flows rectilinearly through the test chamber T, the same partial vacuum exists in it so that it has to be closed off air-tight. During the test the chamber can be entered only through a double set of doors. Eiffel[1] built another larger wind-tunnel installation in 1914, which, however, does not differ fundamentally from the one just described. It uses a different type of blower and a long diffuser or gradually widening channel instead of the intake nozzle of Fig. 218. The diameter of the jet is nearly 7 ft, and the maximum wind velocity is about 130 ft/sec.

Another installation utilizing a free jet is the one in the Aerodynamic Laboratory at Vienna built between 1911 and 1914 by

[1] EIFFEL, G., "New Researches on the Resistance of Air and Aviation" (French), Paris, 1914.

R. Knoller.   Besides having a vertical jet, it differs from Eiffel's tunnel mainly in the construction of the intake nozzle.

146   Modern English Tunnels.—The tunnel of 1910 (Fig. 217)

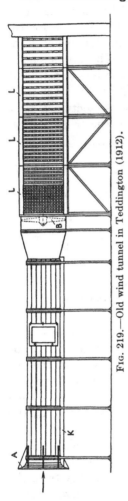

FIG. 219.—Old wind tunnel in Teddington (1912).

had the disadvantage of pulsations in the air velocity. Elaborate researches[1] into the cause of this phenomenon led to the construction of another type which is used to a great extent in England[2] (Fig. 219). The entire structure is set up in a large hall about 6 ft above the floor. The channel has a square cross section of 4 by 4 ft and is 25 ft long. On the intake side, it is rounded (A), and the blower B is built in a somewhat wider section. Up to this point the construction is very much similar to the one of Riabouchinsky. The improvement on this construction, however, consists in the fact that the air is not blown directly into the room but into a long channel L having a great number of small openings from which the air escapes at a reasonably slow velocity. Because of this, the state of turbulence of the air in the hall outside the wind channel is considerably less. Without the muffler L the non-uniformity of the velocity in the channel would amount to ±5 per cent, whereas the muffler reduced it to ±1 per cent. The models can be observed through a glass door placed at about 15 ft from the entrance of the channel.

In 1919 another larger wind tunnel of 7- by 7-ft cross section was built on the same general principles.[3] In this construction the wind channel itself was made to widen out gradually behind

[1] BAIRSTOW, L., and H. BOOTH, An Investigation into the Steadiness of Wind Channels, *Rept. Nat. Adv. Comm. Aeronautics,* 1912–1913, p. 48, London, 1913.

[2] BAIRSTOW, L., J. H. HYDE, and H. BOOTH, The New Four-foot Wind Tunnel, *Rept. Nat. Adv. Comm. Aeronautics,* 1912–1913, p. 59, London, 1913.

[3] *Rept. Nat. Adv. Comm. Aeronautics,* vol. 1, p. 151, 1918–1919; and vol. 1, p. 283, 1922–1923.

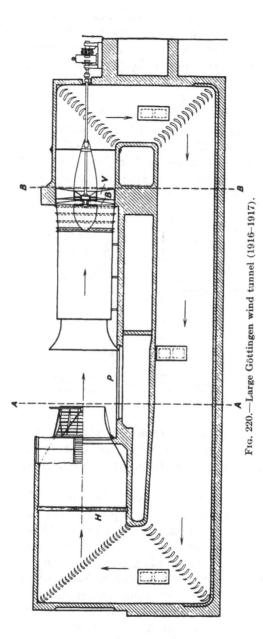

Fig. 220.—Large Göttingen wind tunnel (1916–1917).

the models. This makes the velocities in the blower somewhat smaller and also recovers part of the kinetic energy of the air at the model by converting it into pressure. Instead of using a muffler of the construction of Fig. 219, the air was forced by the blower through a very coarse sieve of brick work into another room, from which it entered into the main hall again through another part of the same sieve.

147. The Large Wind Tunnel in Göttingen.—For the large installation built in Göttingen in 1916–1917 by Prandtl, a free-jet construction was chosen rather than a closed channel. The older Göttingen tunnel (Fig. 216) having a closed channel had shown that in many cases the walls were responsible for errors in

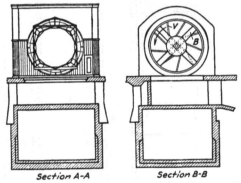

Section A-A   Section B-B

FIG. 221.—Sections of Göttingen tunnel (1916–1917).

the test results. The reason for such errors, as was discussed before, is that for large test objects the constriction of the jet by the solid walls increases the wind velocity at the model.

The construction differed from Eiffel's tunnel in so far as the free jet was made at atmospheric pressure and not at a partial vacuum. Because of this, the test place could be kept entirely open, so that its accessibility was not impeded in any way. Eiffel's tunnel has the disadvantage that the air discharge of the blower into the free atmosphere becomes very turbulent and has to be smoothed out again by special means. This was avoided in the Göttingen construction by inserting a diffuser between the model and the blower. As shown in Figs. 220 and 221, the air discharging from the blower B first passes through some stationary blades V which serve the purpose of annihilating the rotation of the stream imparted to it by the propeller. The air stream is then deflected downward through an angle of 90 deg. by means of a series of blades of special construction. After another turn of 90 deg., it enters into a square cross-sectional channel in the basement of the building. From there it finally enters through a rectifier or sieve *H* into the nozzle chamber having a cross section of 15 by 15 ft. It is then accelerated into the nozzle, which is of circular cross section with a smallest diameter of $7\frac{1}{2}$ ft, causing the velocity to be five times as great. From the nozzle it flows through the model space as a free jet.

The main advantage of this construction lies in the fact that the smoothing out of the air by means of a sieve takes place in the largest channel cross section, which results not only in a smaller power loss but also in a much more efficient smoothing process. As was stated before, the velocity in the jet is five times as great as the velocity in the nozzle chamber so that an air particle in the jet has twenty-five times as much kinetic energy as one at the sieve. After smoothing out the air stream, the deviations from the ideal state have only one twenty-fifth of the kinetic energy of the jet so that of each air particle $^{24}\!\!/_{25}$ths of the energy is transmitted to the nozzle. Even if it were possible to reduce in the sieve the error in the kinetic energy to only 50 per cent, this would result in an energy variation of 2 per cent in the jet itself, *i.e.*, a velocity variation in the direction of the jet of only 1 per cent.

148. Wind Tunnels in Other Countries.—A wind tunnel of considerably larger dimensions was built in 1927 at Langley Field, Va., the center of aeronautic research in the United States.[1] The channel is of the closed type shown schematically in Fig. 222. The jet has a cross section of over 300 sq ft. A still larger wind tunnel of similar construction was completed in 1932 having a jet cross section of about 1300 sq ft. Another wind tunnel worthy of mention in the United States is the one designed by von Kármán in Pasadena, Calif., which is remarkable for its freedom from turbulence and its great over-all efficiency.

It is not necessary to go into details regarding installations in other countries since the constructions are all more or less of the

---

[1] WEICK, F. E., and D. H. WOOD, The Twenty-foot Propeller Research Tunnel of the N. A. C. A., *Rept*. 300, *Nat. Adv. Comm. Aeronautics*, Washington, 1928.

described types.   Regarding the wind tunnels in Japan[1] and in Russia,[2] reference is made to the literature.

The development in the construction of wind tunnels has been guided by the quest for two properties: first, to obtain an air

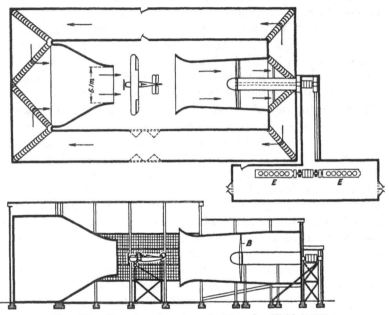

FIG. 222.—New American tunnel at Langley Field (1928).

stream of great uniformity;[3] and, second, to carry out the tests with high Reynolds' numbers in order to obtain dynamic similarity with the actual conditions.   This second desideratum has led to the very large dimensions now in use, especially in England and the United States, and to the very large wind velocities (Göttingen 180 ft/sec, Moscow 260 ft/sec).   A completely

[1] The Resistance of Airship Models Measured in the Wind Tunnels of Japan (English), *Rept. Aero. Research Inst., Tokio Imperial Univ.*, No. 15, March, 1926, The Wind-tunnel Committee of the Aeronautical Council of Japan.

[2] OZEROFF, G. A., The Central Aero-hydrodynamical Institute, *U. S. S. R. Sci. Tech. Dept., No. 183, Supreme Council of National Economy* (Russian), Moscow, 1927.

[3] WIESELSBERGER, C., On the Improvement of the Flow in Wind Tunnels (German), paper read before the 108th meeting of the *Japan. Soc. Mech. Eng.*, 1925.

different method of attack was proposed by Munk,[1] who used moderate dimensions and velocities but reached high Reynolds' numbers by making the kinematic viscosity $\nu = \mu/\rho$ very small. This was done by compressing the air to about twenty times atmospheric pressure which diminishes the kinematic viscosity to about one-twentieth its original value, so that the Reynolds' number becomes twenty times as large. The method naturally involves great structural difficulties since the entire installation has to be able to withstand an internal pressure of 20 atmospheres. The outside shell is made of steel plates of 2-in. thickness. A sketch of the construction actually carried out in the United States is shown in Fig. 223. The entrance to the tunnel is through the door *T*. The various forces are measured auto-

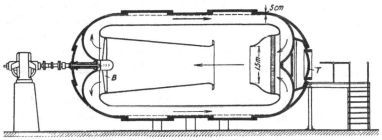

Fig. 223.—High-pressure wind tunnel designed by M. Munk.

matically by recording instruments or by small servo-motors electrically controlled from the outside to change the load on the various balances, which can be observed from the outside through small windows. Recently another larger tunnel on the same principle was built in England.

149. Suspension of the Models and Measurement of the Forces.—Of importance for the suspension of the models in the air stream is the knowledge of the forces exerted by the stream on the suspension wires or struts. It is evident that these forces have to be made as small as possible. If a suspension by means of a strut is employed, the strut is preferably attached to the model in a location where the air would be in a turbulent state without the strut (with a sphere, for instance, on the rear side). This reduces the errors introduced by the strut to a minimum.

It is further desirable that the suspension be made such that

[1] MUNK, M., and E. W. MILLER, The Variable-density Wind Tunnel of the N. A. C. A., *Repts.* 227 and 228, *Nat. Adv. Comm. Aeronautics*, Washington, 1925.

the various components of the air forces on the model can be measured conveniently. A free body in space has six degrees of freedom so that six quantities have to be measured (Fig. 224):

1. The drag D.
2. The lift L.
3. The lateral force Y.
4. The pitching moment M, *i.e.*, the moment round Y as axis.
5. The rolling moment L, *i.e.*, the moment round D as axis.
6. The yawing moment N, *i.e.*, the moment round L as axis.

In most cases, however, the object under test is symmetrical and is situated symmetrically with respect to the flow: for

FIG. 224.—Forces acting on airplane.

instance, an airplane in forward flight. In such cases, the resultant air force lies in the plane of symmetry and consequently is determined by three components. The lateral force Y as well as the rolling and yawing moments is zero, so that it is necessary to measure only the drag, the lift, and the pitching moment.

The construction of the various balances for measuring these forces differs in various countries. They are discussed in a fairly extended manner in the literature quoted before on the various wind tunnels so that it is not necessary to dwell upon them in this book.

Only two special constructions will be considered here, namely, the three-component balance at Göttingen and the balance employed by Eiffel. The suspension in Göttingen is by means of thin wires, whereas Eiffel employs a strut of streamlined shape.

150. The Three-component Balance in Göttingen.—The object under test, in this case an airfoil (Fig. 225), is attached by wires in the three points a, b, and c to various points on the arms L_1 and L_2 of the two balances. From the point a two wires in V-shape connect to the balance L_1; similarly two wires connect c to L_2. The point b, however, is connected with a single vertical wire to L_1. The V-wires at a and c are for the purpose of preventing a side sway of the model.

When lift forces are to be measured, the models are usually suspended upside down so that the lift is pointing downward and can be measured by a wire in tension from the model upward. However, with negative angles of attack the models sometimes are subjected to negative lift forces, and, in order to take these, it is necessary to give the measuring wires some initial tension by means of small weights W_1 and W_2. In order not to subject the test model itself to stresses, these compensating weights have been attached to the points a, b. First the various balances are put in equilibrium when no wind is blowing, in order to take care of the effect of the compensating weights as well as of the weight of the model itself. If then the model is subjected to the air stream causing a lift, this lift is measured by the sum of the forces exerted on L_1 and L_2. The

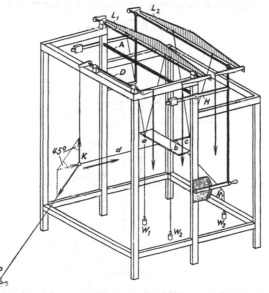

Fig. 225.—Three-component aerodynamic balance at Göttingen.

pitching moment is determined by the force measured by L_2. For the measurement of the drag a wire is stretched from the middle of the model to the point K whence one wire goes up vertically to the drag balance D and another wire under 45 deg. down to a fixed point P. A decomposition of the drag force in these directions causes it to be measured on the balance D, as is shown by the force diagram of Fig. 225. The drag wire is also given initial tension by means of a small counterweight W_3. The drag of the suspension wires is measured separately by replacing the model by another object of very simple shape whose resistance can be calculated. The drag of the original model finally is the difference between the measured drag of it and the drag of the suspension.   In all airfoil and airplane measurements it is of importance to know the air forces for various angles of attack, hence it is convenient to have an apparatus for varying the angle of attack without changing the suspension of the model itself.   In order to do this, the balance L_2 can be lowered or raised round the axis A by means of a lever H.   This causes the model to turn about the axis ab so that various angles of attack can be obtained.   To be sure that the suspension wires at c remain exactly vertical after rotation, it is necessary that the horizontal distance between ab and c be exactly equal to the

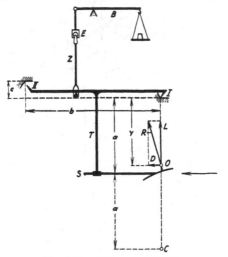

FIG. 226.—Balance of Eiffel.

horizontal distance between the balance L_2 and the axis A. Further details regarding the balance, its calibration, and sources of error are described in the literature.[1]

151. The Aerodynamic Balance of Eiffel.[2]—A T-shaped balancing arm T has knife-edges at its ends I and II (Fig. 226). The test model is attached to T by means of the streamlined strut S.   The arm Z of the balance can be lengthened or shortened by means of the eccentric E so that the horizontal arm of T can be raised or lowered and either one of the two knife-edges I or II can be made to touch its support.   In the first case the balance measures the moment about I and in the second case the moment about II.

[1] *Göttinger Ergebnisse* (German), vol. 1, Munich, 1921.
[2] See footnote, p. 255.

Let the resultant air force exerted on the model be given in magnitude and position by the vector R. This vector R is decomposed into a lift vector L and a drag vector D in the figure, both passing through the point O which is taken to be vertical under the knife-edge I. The moments about I and II are respectively,

$$M_I = -Dy,$$
$$M_{II} = Lb - D(y - c).$$

A third measurement is needed. For this we take the moment round the hypothetical point C, which is the mirror image of I about S as an axis. The measurement about C is carried out by inverting the model and measuring the moments about I, giving

$$M_C = D(2a - y).$$

From these three equations the three quantities D, L, and y can be calculated to be

$$\text{Drag} = D = \frac{M_C - M_I}{2a},$$

$$\text{Lift} = L = \frac{1}{b}(M_{II} - M_I) + \frac{c}{2ab}(M_C - M_I),$$

$$y = -2a\frac{M_I}{M_C - M_I}.$$

D. VISUALIZING FLOW PHENOMENA

152. Fundamental Difficulties.—Since the liquids or gases studied are always homogeneous, and since consequently the individual particles in them cannot be distinguished from each other, it is fundamentally impossible to observe the motion in a liquid or gas without using special means.

One way of approaching the problem is to measure the pressure and the velocity at many points of the flow by means of Pitot tubes or similar apparatus. This, however, gives only an average value of the velocity in space as well as in time, and in general the method is too rough to give a complete picture of the entire flow.

The only manner in which a flow phenomenon can be made visible is by inserting very small particles into the fluid which distinguish the elements from each other without changing the density or other properties of the fluid. For a liquid it is usual to mix certain parts of it with a dye, taking care that the density of the colored liquid is not different from that of the uncolored liquid. For the flow in air, a convenient method is to mix smoke with certain parts of the air stream, but care has to be taken that in feeding in the smoke no additional velocity is imparted to the general stream. Another fairly simple way to obtain a rough picture of the flow of gas (air) is by the use of very thin and light threads of silk. For a steady-state flow these threads show the direction of the velocity at the points where they are located. A fairly complete picture of the flow can be obtained by scanning the field with such a thread.

153. Mixing Smoke in Air Streams.—The usual manner of feeding smoke into a gas or air stream is by means of a number of nozzles which are held stationary in the stream. Figures 34 and 35, Plate 14, show photographs taken in this manner. Several good methods of producing smoke exist, of which the most convenient one consists of permitting air saturated with hydrochloric acid to come into contact with the fumes of ammonia. This leads to a white, fog-like smoke which shows a good contrast against a black background. Other methods for producing smoke photographs are described in the literature.[1]

Vortex rings especially are suitable objects for being shown by means of the smoke method. A box closed on all sides has on one side a taut rubber membrane, and on the opposite side it has a circular hole. The box is filled with smoke and then a light tap is given on the membrane. A well-formed smoky vortex ring escapes from the opening and remains intact for a long time in still air since such a vortex structure is in a very stable state of equilibrium. By tapping the membrane a number of times in succession several smoke rings can be produced and the mutual reaction of them can be demonstrated.

154. Motions in the Boundary Layer.—In an investigation of the flow in the wake of a body where there is considerable turbulence, the smoke method is unsuitable since the various smoke threads will become completely mixed with each other. In order to avoid this difficulty, Riabouchinsky[2] used the following

---

[1] *Göttinger Ergebnisse*, vol. 2, Munich, 1923. MAREY, On the Movements of Air Flowing round Various Obstacles (French), *Compt. rend.*, vol. 131, p. 160, 1900; Changes in the Direction and Velocity of an Air Stream (French), *Compt. rend.*, vol. 131, p. 1291, 1901.

[2] RIABOUCHINSKY, D., *Bull. inst. aéro. Koutchino* (French), vol. 3, p. 59, Moscow, 1909.

method for the investigation of a two-dimensional flow: The models were cylinders of various cross sections mounted with their bases on very thin blackened steel plates on which a light-yellow powder (lycopodium) had been deposited. The model was then subjected to a horizontal air stream parallel to the plate, and the plate at the same time was put into vibration by means of light hammer taps. The picture of the powder on the black background then indicated the streamline pattern. A serious objection to this method is that the air flow is recorded not in the free air but rather in the region of the boundary layer on the plate.

The same objection can be raised against another method for visualizing air flows, originated by Fales.[1] He wanted to investigate the angle of attack at which the flow round an airfoil would break away. His airfoil model was of white color and in the middle a disk was attached to it perpendicularly. Both the airfoil and the disk were coated with a suspension of lampblack in kerosene. Owing to the action of the wind, the kerosene flowed along the model as well as along the disk and showed the flow as white lines on a black background.

The various methods for visualizing flow phenomena in water can be divided into two groups, depending on whether it is desired to study the flow in the interior of the fluid or only on its surface.

155. Three-dimensional Fluid Motions.—In the case of a three-dimensional fluid motion, it is necessary to study the flow not only on the surface but also in the interior of the fluid. The flow can be shown in such cases by inserting colored water of the same specific gravity through a number of nozzles, but care has to be taken that the velocity of efflux at the nozzle is equal to the velocity of the surrounding fluid. For water a suitable coloring is potassium permanganate or certain kinds of anilin dyes dissolved in a small amount of alcohol and then distributed into a large volume of water. If great accuracy is necessary, these dyes can be given the same specific gravity as water by adding to them other liquids of suitable specific gravity before mixing them with the water. If it is intended to make photographs of the flow instead of making visual observations only, the choice of color is determined by somewhat different considerations. Then

---

[1] FALES, E. N., Visible Study of Flow, *McCook Field Report, Serial* No. 2635, Published by the Chief of Air Service, Washington, 1926.

it is desirable to choose a color that would make hardly any impression on the photographic plate, for instance, certain red dyes usually employed for coloring the glass in photographic darkrooms.   In cases when there is a dark background, skimmed milk is a suitable coloring agent.   Another method of showing the flow is to use a suspension of extremely small aluminum particles in the water, which on account of their size remain floating for a very long time.   Such a suspension can be made by first wetting the aluminum powder with alcohol and then pouring it into a bottle of water which is shaken violently.   An advantage of this procedure over the color method is that the aluminum suspension is still capable of showing details of the flow in a turbulent region where a colored jet becomes completely mixed up.[1]

A method of showing motions in the boundary layer near to a body immersed in water consists of coating the body with a color which can be dissolved easily in water or of painting it with condensed milk.   In this connection it is of interest to mention the method of Thoma,[2] who uses a precipitate which is formed by the chemical action of the air on a substance painted on the surface of the body and which evaporates and comes into contact with the air by diffusion.   Since diffusion is subject to the same laws as the change in velocity due to internal friction, this method gives a coloring of the boundary layers and of the turbulent regions.   Thoma wrapped the body in blotting paper saturated with hydrochloric acid.   The air was mixed with ammonia vapor, causing a white fog in the region of diffusion.

Another method for making boundary-layer motion visible is that of Simmons and Dewey.[3]   The model in this case is painted with titanium tetrachloride ($TiCl_4$), which evaporates as a white fog.

It is also possible to mix the entire fluid with a suspension of very small aluminum particles; the concentration of the particles has to be made rather small in order to make it possible to look sufficiently far into the fluid.   In case only certain path

---

[1] ERMISCH, H., Flow Pattern and Pressure Distribution on Various. Objects as a Function of the Reynolds' Number (German), *Abhandl. Aero. Inst., Tech. Hochschule Aachen,* vol. 6, p. 21, Berlin, 1927.

[2] THOMA, H., Highly Efficient Boilers (German), Berlin, 1921.

[3] SIMMONS, L. F. G., and N. S. DEWEY, Wind-tunnel Experiments with Circular Disks, *Repts. and Mem. Nat. Adv. Comm. Aeronautics,* No. 1334, London, 1931; Photographic Records of Flow in the Boundary Layer, *Repts. and Mem. Nat. Adv. Comm. Aeronautics,* No. 1335, London, 1931.

lines are wanted, it is practical to take a smaller number of larger suspended particles. A method for the making of spheres of the same specific gravity as water for this purpose has been given by Marey.[1] These spheres are made of a mixture of wax of specific gravity 0.96 and rosin of specific gravity 1.07. The spheres so obtained are silvered like pills in the pharmacy. They are made a trifle heavier than water so that they sink down slowly; then some salt is mixed with the water until the point of perfect equilibrium is reached. Another method which is often used in England consists of injecting oil into the fluid by means of an atomizer. The resulting droplets are then illuminated sharply by means of a thin sheet of light.[2] Satisfactory results have been obtained by drops made of a mixture of olive oil and nitrobenzol or of a mixture of tetracarbonchloride and xylol.

156. Two-dimensional Fluid Motions.—In case the flow is a two-dimensional one, it can be visualized in a much simpler manner since it is usually necessary to observe only the motions on the water surface in a tank. As an example, let the model be a cylinder of which the base sits on the bottom of the tank and of which the top just protrudes from the surface of the water. In such a case, the flow is completely two dimensional, *i.e.*, the same in all planes parallel to the surface. Care has to be taken that no capillary effects come in, by keeping the surface of the water meticulously clean. Even the dipping in of a clean hand, or a few hours' contact of the water surface with the atmosphere, makes the surface useless for this purpose. To be certain that the condition of the surface is satisfactory, the following simple test can be made. Sprinkle some aluminum powder on the water and then blow vertically down on it with the mouth. This spreads the aluminum particles in all directions and clears a circular area of the surface. If, after the blowing, the aluminum particles remain where they are, the surface is clean; if, however, the circle closes up by itself, the surface is contaminated and has to be renewed. This can be done in the simplest way by using an

[1] MAREY, Experimental Hydrodynamics (French), *Compt. rend.*, vol. 116, p. 913, 1893.

[2] EDEN, C. G., Investigation by Visual and Photographic Methods of the Flow past Plates and Models, *Rept. Nat. Adv. Comm. Aeronautics*, 1911–1912, p. 97, London, 1912; RELF, E. E., Photographic Investigations of the Flow round a Model Airfoil, *Rept. Nat. Adv. Comm. Aeronautics*, 1912–1913, p. 133, London, 1913.

overflow.   Given a clean surface the flow can be shown very satisfactorily by means of aluminum powder or lycopodium powder.   The first experiments of this sort were made in 1900 by Ahlborn.[1]   Later Rubach[2] photographed the motion of a vortex pair behind cylindrical bodies in motion.   Most of the photographs shown at the end of this book have been made in this manner by the author.   To prevent the aluminum particles from running away from the model under the influence of the capillary angle between the water surface and the model it is helpful to coat the latter with a thin layer of paraffin.   By means of this procedure it is possible to prevent the capillary action so that the fluid surface remains completely horizontal at the model.   It is even possible to create a negative capillary angle by lowering the model somewhat, which may be useful for showing the history of the boundary-layer particles.   Under the influence of a negative capillary angle the aluminum particles are crowded round the model, and after a short time of motion it can be seen clearly where these boundary-layer particles move (see Fig. 7, Plate 4).

Another method of visualizing the motion of water is due to Prandtl (1904).[3]   He suspended very small flakes of mica in the water.   Certain regular motions, especially vortices, could be seen clearly because a great number of these flakes then had the same orientation.

A serious disadvantage of the method of observing the surface of the fluid is that at relatively small velocities capillary waves are formed.   For water this critical velocity is about 10 in./sec. A circular cylinder moving through the water with a certain velocity shows approximately twice this velocity at some local points so that it cannot be moved at a speed greater than 5 in./sec if capillary waves are to be avoided.

If greater velocities are desired, it is necessary to move the model entirely under water and to photograph the motion not of the top surface but of a plane parallel to it in the water.   The technical difficulties of such a procedure, however, are great. The best method is to mix the water with drops of a mixture

[1] AHLBORN, F., On the Mechanism of Hydrodynamic Drag (German), *Abhandl. Gebiete Naturwiss.*, vol. 17, Hamburg, 1902; or *Jahrb. Schiffbautechn. Gesellsch.*, 1904, 1905, and 1909.

[2] RUBACH, H., On the Generation and Motion of the Vortex Pair behind Cylindrical Bodies (German), Dissertation, Göttingen, 1914; or *Forschungsarbeiten V. D. I.*, vol. 185, 1916.

[3] See footnote, p. 58.

of olive oil and nitrobenzol and then illuminate only a plane sheet of the water under the surface. Because the oil drops are illuminated under an angle of 90 deg. with respect to the direction of motion they are very clearly visible, whereas the drops in the fluid above the illuminated surface do not impede vision to a great extent. This latter property makes oil drops more practical than the spheres of wax and rosin discussed before.

157. Advantage of Photographs over Visual Observations.— Aside from the fact that photographs are far more convincing to an outsider than a mere description of the observations of an experimenter, a photograhic record of a flow will disclose many facts which cannot be obtained by visual observation. It has been shown in Art. 37, "Fundamentals,"[1] that the shape of the streamlines depends very much on the choice of the coordinate system. For instance, if an airfoil model is moved through water the streamlines obtained by photographing the aluminum powder with the camera standing still with respect to the water are different from those taken with the camera at rest with respect to the model. In the first case the streamline picture has the appearance of Fig. 52, Plate 21, whereas in the second case Fig. 50, Plate 20, is obtained. An observer without special training sees only the pictures of the second kind since the eye has the tendency to follow the model in its motion.

Moreover, the element of time is very important. Whereas a photograph can be studied at leisure, and many details can be found in this manner, the visual observer has to digest all his information in a few seconds. Another advantage of photographic records, especially moving pictures, is that they can be shown repeatedly, which is very instructive.

158. Streamlines and Path Lines.—The photographs obtained do not give the streamlines with mathematical accuracy. The picture consists of short stretches of curve of various length due to the motion of the individual aluminum particles during the time of exposure of the plate. Therefore these little stretches are parts of path lines. On the other hand, during a short interval the path lines and streamlines have the same tangent so that the picture made of the various stretches of path line put together gives the appearance of a field of streamlines. Therefore streamlines appear the more exact the shorter the time of exposure is.

[1] See footnote, p. 3.

If the velocity field is independent of time (steady flow), it was seen in Art. 35, "Fundamentals,"[1] that streamlines and path lines are identical.   In such a case, streamlines could be obtained also with a *long* exposure (Fig. 50, Plate 20).   For non-steady motions (Fig. 52, Plate 21), the streamline picture changes with the time; only short exposures show approximately the instantaneous stream geometry, whereas a long exposure gives path lines which in general give a very irregular and useless picture. An analysis of a number of consecutive streamline pictures follows the procedure of Euler, whereas an analysis of path lines requires the use of Lagrange's method (see Art. 34, "Fundamentals"[1]). In the case of three-dimensional flow phenomena it is useful and sometimes necessary to make stereoscopic photographs.

159. Slow and Fast Moving Pictures.—In case the actual phenomena occur at a very fast rate, it is useful to take a great number of exposures during a short time and reproduce them later at a slower rate, which has the effect of slowing up the motion.   With the usual motion-picture camera 16 to 20 pictures per second are made and the reproduction on the screen is at the same rate.   With special cameras, up to 2,000 pictures per second can be taken.[2]

For instance, if a phenomenon has been photographed 320 times per second and is reproduced at the rate of 16 pictures per second, the motion appears twenty times as slow as in the actual case.   For motions of great rapidity for which this is not sufficient (flying bullets, explosions, cavitation), the number of exposures per second has to be even greater.   For this purpose ballistic engineers have constructed cameras with intermittent illumination by means of electric sparks, which are capable of taking 40,000 pictures per second.[3]   By the use of several lenses

[1] See footnote, p. 3.

[2] THUN, R., Application and Theory of the "Time Stretcher" (German), *Z. V. D. I.*, p. 1353, 1926.

[3] The upper limit of the number of pictures per second is not caused by the frequency of the sparks, which can easily be increased to 100,000, but is rather due to the strength of the film which moves with very great velocity in order to obtain sufficiently large pictures.   For instance, for 6,000 pictures per second at ½-in. width each, the film has to move at a speed of 250 ft/sec.   See paper by Terazawa, Kinematographic Study of Aeronautics (English), *Rept. Aero. Research Inst., Tokio Imperial Univ.*, vol. 1, p. 8, 1924.

the frequency of the pictures can be increased still further, up to about 300,000 per second.[1]

The opposite procedure of taking a smaller number of pictures than are reproduced later has also been used.[2] This is useful for reproducing very slow motions; for instance, if the motion of a cloud during 50 min has to be shown, it is of advantage to take one picture every 5 sec and to reproduce it at the usual rate of 20 pictures per second. The whole phenomenon then takes place in $\frac{1}{2}$ min, which reveals the characteristic motions of the cloud in a very striking manner.

160. Long-exposure Moving Pictures.—Since in a usual moving picture the exposure is about $\frac{1}{40}$ sec, a single picture does not show anything about the state of flow. For instance, a picture taken of the water tank with aluminum powder will show the various aluminum particles as dots only. If the cranking of the moving camera is slowed down so that only two exposures per second are made, the illumination is about $\frac{1}{4}$ sec for each picture. Then the particles show as short lines so that one single photograph in itself gives a clear idea of the flow. If a print were made on transparent paper and several consecutive pictures were put on top of each other, the various images of the same aluminum particle would form a dashed line. The various dashes show the exposures, whereas the gaps between the dashes are due to the closing of the lens during the transportation of the film to its next position. The lengths of the dashes as well as of the gaps are a measure of the instantaneous velocity of the particular particle under consideration.

In order to shorten the time of non-exposure and thus to improve the possibility of interpolation, Prandtl has suggested a modification of the usual motion-picture camera[3] in which, by means of the insertion of a "maltese cross" into the drive, the time necessary for transportation of the film is reduced to about one-twelfth of the time of exposure. Figures 7, 8, and 9 on Plates 4, 5, and 6 show a number of films taken by the author with this apparatus.

[1] CRANZ, C., and H. SCHARDIN, Kinematography on a Non-moving Film with Extremely High Frequency (German), *Z. Physik*, vol. 56, p. 147, 1929.

[2] NEUMANN, H., Time-condensing Pictures (German), *Kinotechnik*, vol. 9, p. 173, 1927.

[3] PRANDTL, L., and O. TIETJENS, Kinematographic Flow Pictures (German), *Naturwissenschaften*, vol. 13, p. 1050, 1925.

It is mentioned in passing that it has been found useful in many cases to photograph not only the flow but also a clock, a scale, etc., in order to record the time, velocities, and distances immediately on the film itself.

161. Technical Details.—Since it is desired to obtain the greatest possible contrast in the flow photographs, it is of importance to make the background as black as possible. For this purpose, black velvet is very useful since it reflects less than 1 per cent of the incident light, whereas from dull-black metal surfaces or black paper about 10 per cent of the light is reflected. In order to get the maximum contrast in the pictures, it is generally better to use intense illumination and good lenses than sensitive plates or films. The best plates or films available are relatively non-sensitive ones, since with them the curve giving the relation of the blackening as a function of the intensity of illumination is much steeper than with hypersensitive plates or films. Another advantage of the less sensitive plates is that the grain is much finer. If the sensitivity of an ordinary fine-grained plate is not sufficient for the purpose in hand, it can be raised about thirty or forty times by suitable baths without increasing the size of grain.[1] If there exists danger of underexposure, it is well to raise the sensitivity threshold of the film by previously exposing it to a very weak source of light in the dark room, such that the plate remains clear but is just on the point of blackening.

The choice of lens is not of particular importance since very little depth focusing is required in the picture. Any powerful lens of short focal length serves the purpose, for instance, a well-corrected doubly anastigmatic lens.

Regarding illumination, it has to be borne in mind that it is desired to make an impression on the photographic plate and not on the eye. For the usual non-sensitized plates the maximum sensitivity is for light of about $\lambda = 400$ Å. Therefore lamps with much ultra-violet light like mercury-arc or carbon-arc lamps are better suited than the usual incandescents. Among arc lamps the enclosed types are better than the open-arc ones, since for the same watt consumption they give three to four times as much actinic light. For short exposures ($\frac{1}{20}$ sec) flash-light powder

[1] GUILLEMINOT, P., Hypersensitizing and Ultrasensitizing (French), *Rev. franç. de photog. et cinématog.*, vol. 8, No. 181, 1927; SHEPPARD, S. E., Increasing the Sensitivity of Silver Emulsions (German), *Die Photographische Industrie*, p. 1032, 1925.

is useful. In order to increase the length of time of the flash which is necessary for flow photographs, a mixture of magnesia powder and some inert powder is used.

In choosing the developing and fixing baths it is to be remembered that for ordinary snapshots it is desirable to have the various shades of darkness merge into each other gradually, while here sharp contrasts are wanted. Therefore hydrochinon is useful as a developer since it makes very black pictures. Special developers made for titles in moving pictures are also suitable.

If the film is underexposed it is desirable to overdevelop it, so that the unexposed parts start to blacken. This blackening is then removed by means of a reducing bath (Farman reducer) and the film is then intensified in a uranium bath. Though for ordinary pictures the uranium intensifier is capricious, it is very suitable for obtaining negatives of great contrast.

The best paper for making prints is non-sensitive extra-glossy "developing-out" paper, having a steep blackening curve.[1] In order to get a still better gloss the paper may be dried on plate glass, which gives pictures that are very suitable for reproduction in print.

[1] GOLDBERG, E., The Composition of the Photographic Picture (German), vol. I, Halle, 1925.

PLATES

The flow photographs are made at the Kaiser Wilhelm Institute for Flow Research (Göttingen, Germany) with an experimental equipment developed by the author.

PLATE 1.[1]

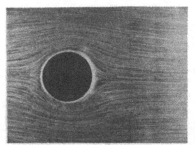

FIG. 1.—Flow round cylinder immediately after starting (potential flow).

FIG. 2.—Backward flow in the boundary layer behind the cylinder; accumulation of boundary layer material.

PLATE 2.

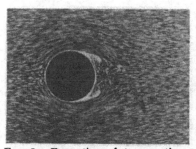

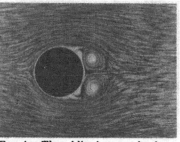

FIG. 3.—Formation of two vortices; flow breaking loose from cylinder.

FIG. 4.—The eddies increase in size.

PLATE 3.

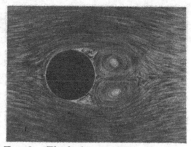

FIG. 5.—The eddies grow still more; finally the picture becomes unsymmetrical and disintegrates.

FIG. 6.—Final picture obtained a long time after starting.

[1] The direction of flow in all photographs is from left to right.

279

PLATE 4.

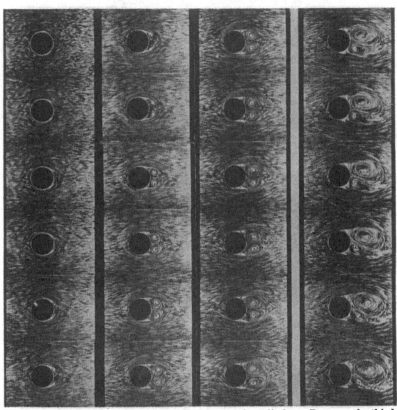

FIG. 7.—Consecutive pictures of the flow round a cylinder. Between the third and fourth vertical column a number of pictures is missing. The fourth column shows the disintegration of the symmetrical vortices ending in a picture like that of Fig. 6.

PLATE 5.

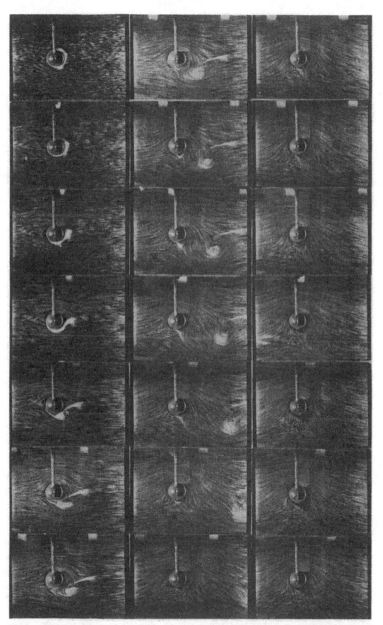

FIG. 8.—Consecutive pictures of the flow round a rotating cylinder starting from rest. The ratio of peripheral velocity of rotation u to the forward velocity v is $u/v = 4$.

PLATE 6.

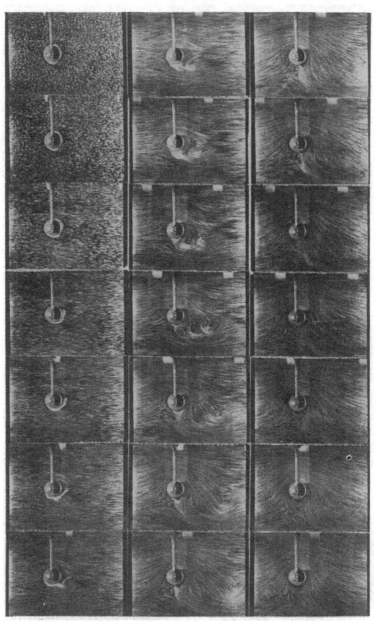

FIG. 9.—The same as Fig. 8, except that $u/v = 6$.

PLATE 7.

FIG. 10.—$u/v = 0$.

FIG. 11.—$u/v = \frac{1}{2}$.

FIG. 12.—$u/v = 1$.

PLATE 8.

FIG. 13.—$u/v = 2$.

FIG. 14.—$u/v = 3$.

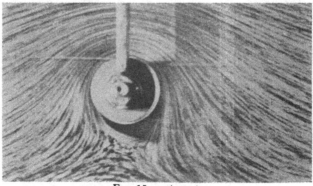

FIG. 15.—$u/v = 4$.

PLATE 9.

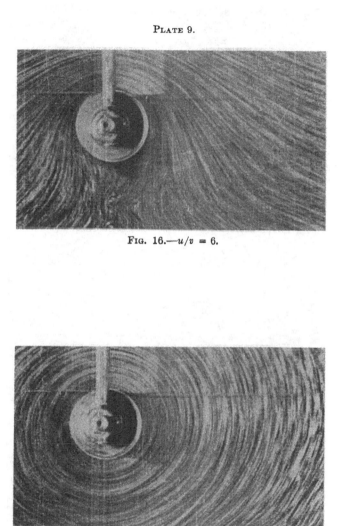

FIG. 16.—u/v = 6.

FIG. 17.—u/v = ∞. This picture was taken by moving the rotating cylinder with the camera through the water, stopping both and immediately afterward exposing the plate.

PLATE 10.

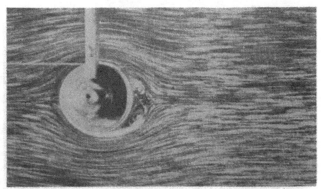

FIG. 18.—$u/v = \frac{1}{2}$.

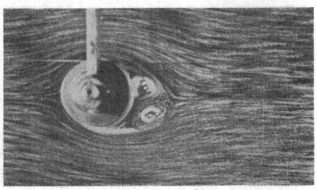

FIG. 19.—$u/v = \frac{1}{2}$.

FIG. 20.—$u/v = \frac{1}{2}$.

FIGS. 18–20.—Consecutive stages of development of the flow for $u/v = \frac{1}{2}$.

PLATE 11.

FIG. 21.—$u/v = 3$.

FIG. 22.—$u/v = 3$.

FIG. 23.—$u/v = 3$.

FIGS. 21–23.—Consecutive stages of development of the flow for $u/v = 3$.

PLATE 12.

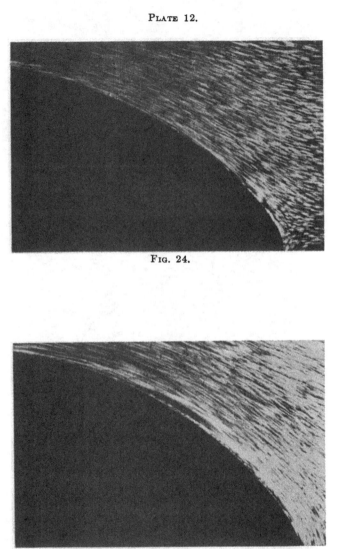

FIG. 24.

FIG. 25.
FIGS. 24–33.—Flow along rear end of blunt body.

PLATE 12.—(*Continued*)

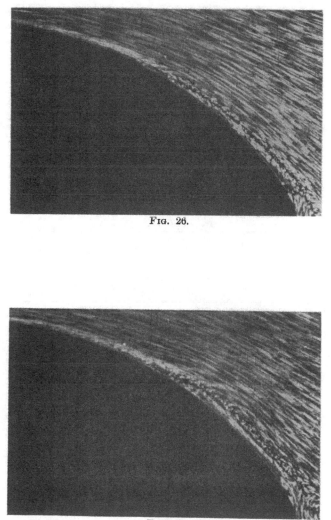

FIG. 26.

FIG. 27.

PLATE 13.

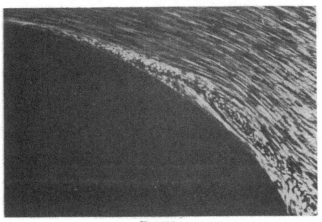

FIG. 28.

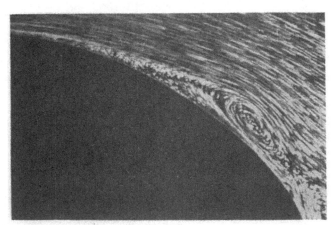

FIG. 29.

Plate 13.—(*Continued*)

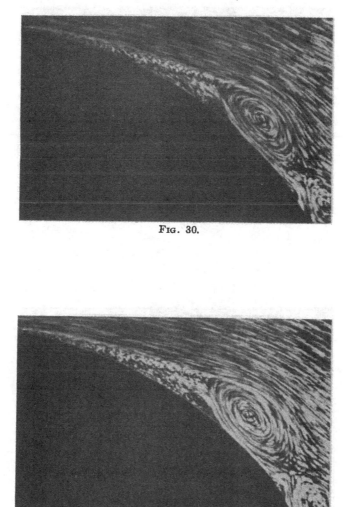

Fig. 30.

Fig. 31.

PLATE 14.

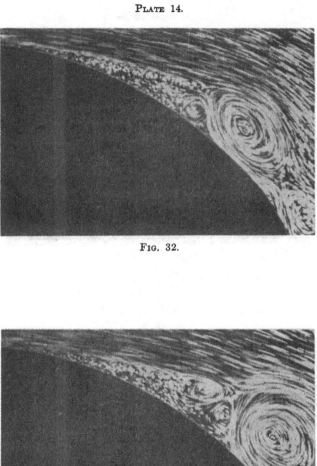

FIG. 32.

FIG. 33.

PLATE 14.—*(Continued)*

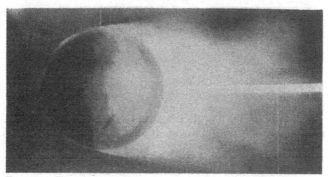

FIG. 34.—Flow round sphere below critical point. (*Wieselsberger.*)

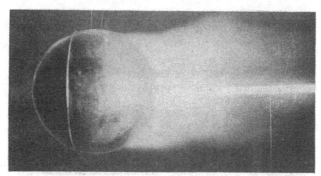

FIG. 35.—Owing to a thin wire ring round the sphere, the flow becomes of the other type with turbulent boundary layer. (*Wieselsberger.*)

PLATE 15.

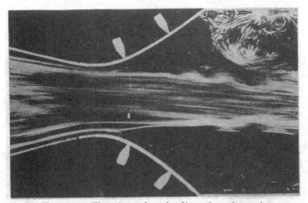

FIG. 36.—Flow in a sharply diverging channel.

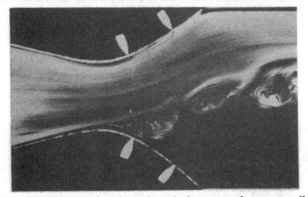

FIG. 37.—The boundary layer is sucked away at the upper wall.

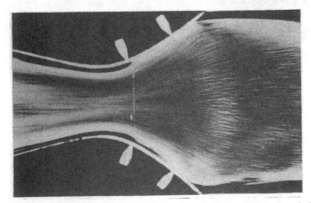

FIG. 38.—The boundary layer is sucked away on both walls; the flow is from left to right.

PLATE 16.

FIG. 39.—Turbulent flow in an open channel; the speed of the camera is about equal to the speed of the water near the walls.

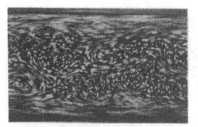

FIG. 40.—As before, but here the camera speed is that of the water in the center.

FIG. 41.—Flow round a knife edge.

PLATE 17.

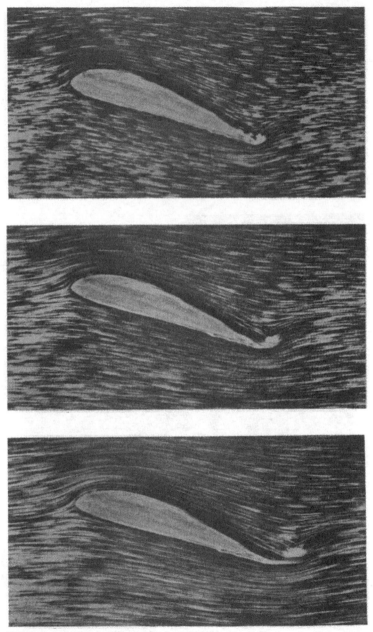

FIGS. 42–44.—Consecutive stages of flow round airfoil starting from rest.

PLATE 18.

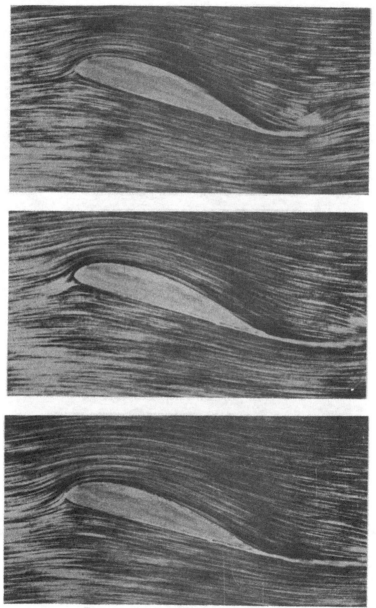

FIGS. 45–47.—Continuation of Figs. 42–44.

PLATE 19.

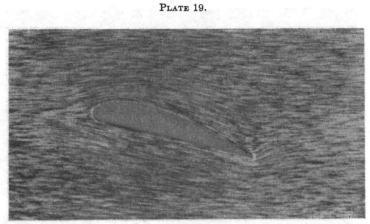

FIG. 48.—Streamlines round an airfoil the very first moment after starting.

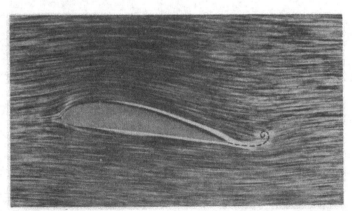

FIG. 49.—Formation of the starting vortex which is washed away with the fluid.

PLATE 20.

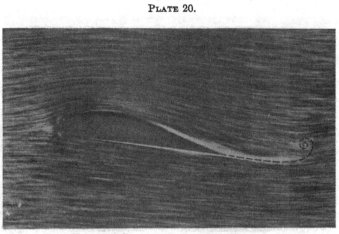

FIG. 50.—Growing of the starting vortex.

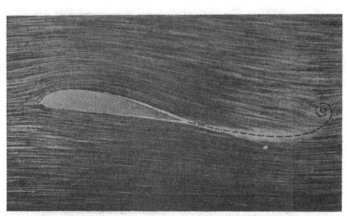

FIG. 51.—Taken somewhat later than Fig. 50.

PLATE 21.

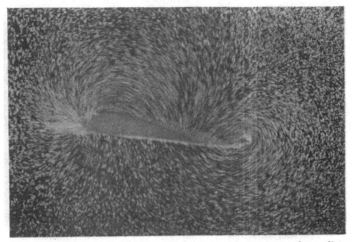

FIG. 52.—Like Fig. 49; but the camera is at rest with respect to the undisturbed fluid.

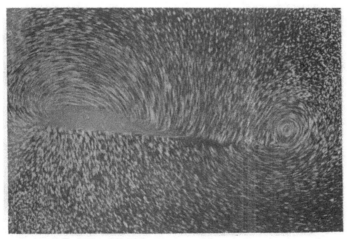

FIG. 53.—Like Fig. 51, but with camera at rest.

PLATE 22.

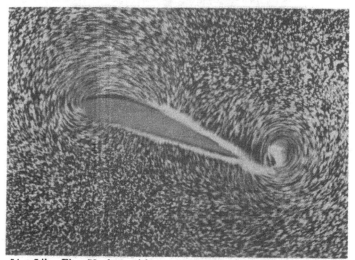

FIG. 54.—Like Fig. 52, but with greater angle of attack and consequently stronger starting vortex. Also shorter exposure of plate.

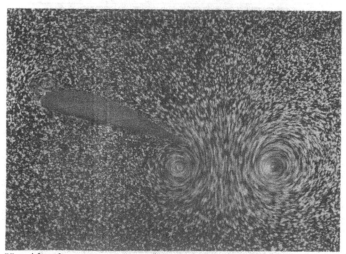

FIG. 55.—After formation of the starting vortex the airfoil was stopped and then the picture was taken.

PLATE 23.

FIG. 56.—$wd/\nu = 0.25$.

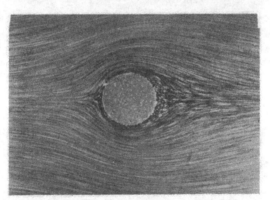

FIG. 57.—$wd/\nu = 1.5$.

FIG. 58.—$wd/\nu = 9$.

FIGS. 56–58.—Flow round cylinder at small Reynolds' numbers.

PLATE 24.

FIG. 59.—Kármán trail; $wd/\nu = 250$. The camera is at rest with respect to the cylinder.

FIG. 60.—Kármán trail; $wd/\nu = 250$. The camera is at rest with respect to the undisturbed fluid.

PLATE 25.

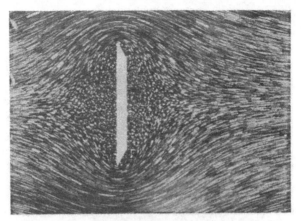

FIG. 61.—$wb/\nu = 0.25$.

FIG. 62.—$wb/\nu = 10$.

FIG. 63.—$wb/\nu = 250$.
FIGS. 61–63.—Flow round sharp plate of width b.

PLATE 26.

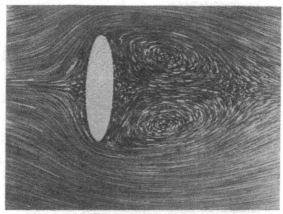

FIG. 64.—$wb/\nu = 10$.

FIG. 65.—$wb/\nu = 80$.

FIG. 66.—$wb/\nu = 250$.
FIGS. 64–66.—Flow round elliptic cylinder with major axis b.

PLATE 27.

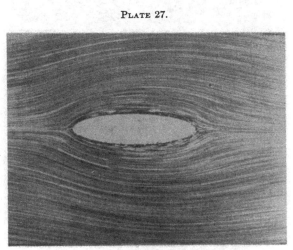

FIG. 67.—Flow round elliptic cylinder of minor axis d at $wd/\nu = 1.3$.

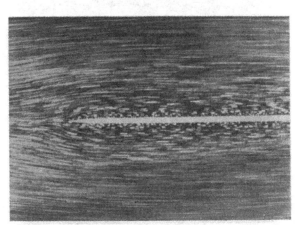

FIG. 68.—Flow round thin plate of length l. $wl/\nu = 3$.

INDEX

INDEX

A

Acceleration, longitudinal, substantial, 2
  resistance due to, 107
Aerodynamic balance, 262
Airfoil, with finite tail angle, 182
  Joukowsky profile, 180
  pressure distribution on, 156
  slotted, 154
  sucking boundary layer from, 155
  theory of finite, 185
  theory of infinite, 158
Airplane, transfer of weight to ground, 186
Anemometer, 239
Angle of attack, 146
Aspect ratio influence of, on drag, 145

B

Backflow in boundary layer, 69
Balance, aerodynamic, 262
Bernoulli's constant, 3
Bernoulli's equation, 3
Biplane, theory of staggered, 213
  theory of unstaggered, 210
Boundary layer, backflow in, 69
  definition of thickness of, 64, 67, 76
  differential equation of, 62
  for flat plate, 66
  order of magnitude of, 61
  sucking of material from, 81
  velocity distribution in laminar, 68
  in turbulent, 70
  visualizing motion in, 266

C

Circulation, definition of, 160
  generation of, 167
Colored line, criterion for turbulence, 34
Conformal mapping, 173
Continuity equation, 1
Convergent flow, 52
Conversion formulas, 206
Correction term for kinetic energy, 24
Critical Reynolds' number, 32

D

Deformation resistance, 88
Differential coefficient, convective, 2
  local, 2
  substantial, 2
Differential equation of Navier-Stokes, 5
Dimensional analysis, 12
Dirichlet's paradox, 108
Discontinuity, surface of, 191
Divergent flow, 52
Downward velocity, induced, 197
Drag, of airfoil, 147
  determination from wake measurements, 126
  with discontinuous potential flow, 110
  of half body, 118
  induced, 198
  measurement of, in natural wind, 250
    in wind tunnel, 251
  momentum theory of, 123
  with potential flow, 104
  self-induced, 210
  starting, 169
Drag coefficient, 92

A CATALOG OF SELECTED
DOVER BOOKS
IN ALL FIELDS OF INTEREST

A CATALOG OF SELECTED DOVER
BOOKS IN ALL FIELDS OF INTEREST

100 BEST-LOVED POEMS, Edited by Philip Smith. "The Passionate Shepherd to His Love," "Shall I compare thee to a summer's day?" "Death, be not proud," "The Raven," "The Road Not Taken," plus works by Blake, Wordsworth, Byron, Shelley, Keats, many others. 96pp. 5³⁄₁₆ x 8¼. 0-486-28553-7

100 SMALL HOUSES OF THE THIRTIES, Brown-Blodgett Company. Exterior photographs and floor plans for 100 charming structures. Illustrations of models accompanied by descriptions of interiors, color schemes, closet space, and other amenities. 200 illustrations. 112pp. 8⅜ x 11. 0-486-44131-8

1000 TURN-OF-THE-CENTURY HOUSES: With Illustrations and Floor Plans, Herbert C. Chivers. Reproduced from a rare edition, this showcase of homes ranges from cottages and bungalows to sprawling mansions. Each house is meticulously illustrated and accompanied by complete floor plans. 256pp. 9⅜ x 12¼.
0-486-45596-3

101 GREAT AMERICAN POEMS, Edited by The American Poetry & Literacy Project. Rich treasury of verse from the 19th and 20th centuries includes works by Edgar Allan Poe, Robert Frost, Walt Whitman, Langston Hughes, Emily Dickinson, T. S. Eliot, other notables. 96pp. 5³⁄₁₆ x 8¼. 0-486-40158-8

101 GREAT SAMURAI PRINTS, Utagawa Kuniyoshi. Kuniyoshi was a master of the warrior woodblock print — and these 18th-century illustrations represent the pinnacle of his craft. Full-color portraits of renowned Japanese samurais pulse with movement, passion, and remarkably fine detail. 112pp. 8⅜ x 11. 0-486-46523-3

ABC OF BALLET, Janet Grosser. Clearly worded, abundantly illustrated little guide defines basic ballet-related terms: arabesque, battement, pas de chat, relevé, sissonne, many others. Pronunciation guide included. Excellent primer. 48pp. 4³⁄₁₆ x 5¾.
0-486-40871-X

ACCESSORIES OF DRESS: An Illustrated Encyclopedia, Katherine Lester and Bess Viola Oerke. Illustrations of hats, veils, wigs, cravats, shawls, shoes, gloves, and other accessories enhance an engaging commentary that reveals the humor and charm of the many-sided story of accessorized apparel. 644 figures and 59 plates. 608pp. 6⅛ x 9¼.
0-486-43378-1

ADVENTURES OF HUCKLEBERRY FINN, Mark Twain. Join Huck and Jim as their boyhood adventures along the Mississippi River lead them into a world of excitement, danger, and self-discovery. Humorous narrative, lyrical descriptions of the Mississippi valley, and memorable characters. 224pp. 5³⁄₁₆ x 8¼. 0-486-28061-6

ALICE STARMORE'S BOOK OF FAIR ISLE KNITTING, Alice Starmore. A noted designer from the region of Scotland's Fair Isle explores the history and techniques of this distinctive, stranded-color knitting style and provides copious illustrated instructions for 14 original knitwear designs. 208pp. 8⅜ x 10⅞. 0-486-47218-3

Browse over 9,000 books at www.doverpublications.com

CATALOG OF DOVER BOOKS

ALICE'S ADVENTURES IN WONDERLAND, Lewis Carroll. Beloved classic about a little girl lost in a topsy-turvy land and her encounters with the White Rabbit, March Hare, Mad Hatter, Cheshire Cat, and other delightfully improbable characters. 42 illustrations by Sir John Tenniel. 96pp. 5³⁄₁₆ x 8¼. 0-486-27543-4

AMERICA'S LIGHTHOUSES: An Illustrated History, Francis Ross Holland. Profusely illustrated fact-filled survey of American lighthouses since 1716. Over 200 stations — East, Gulf, and West coasts, Great Lakes, Hawaii, Alaska, Puerto Rico, the Virgin Islands, and the Mississippi and St. Lawrence Rivers. 240pp. 8 x 10¾. 0-486-25576-X

AN ENCYCLOPEDIA OF THE VIOLIN, Alberto Bachmann. Translated by Frederick H. Martens. Introduction by Eugene Ysaye. First published in 1925, this renowned reference remains unsurpassed as a source of essential information, from construction and evolution to repertoire and technique. Includes a glossary and 73 illustrations. 496pp. 6⅛ x 9¼. 0-486-46618-3

ANIMALS: 1,419 Copyright-Free Illustrations of Mammals, Birds, Fish, Insects, etc., Selected by Jim Harter. Selected for its visual impact and ease of use, this outstanding collection of wood engravings presents over 1,000 species of animals in extremely lifelike poses. Includes mammals, birds, reptiles, amphibians, fish, insects, and other invertebrates. 284pp. 9 x 12. 0-486-23766-4

THE ANNALS, Tacitus. Translated by Alfred John Church and William Jackson Brodribb. This vital chronicle of Imperial Rome, written by the era's great historian, spans A.D. 14-68 and paints incisive psychological portraits of major figures, from Tiberius to Nero. 416pp. 5³⁄₁₆ x 8¼. 0-486-45236-0

ANTIGONE, Sophocles. Filled with passionate speeches and sensitive probing of moral and philosophical issues, this powerful and often-performed Greek drama reveals the grim fate that befalls the children of Oedipus. Footnotes. 64pp. 5³⁄₁₆ x 8 ¼. 0-486-27804-2

ART DECO DECORATIVE PATTERNS IN FULL COLOR, Christian Stoll. Reprinted from a rare 1910 portfolio, 160 sensuous and exotic images depict a breathtaking array of florals, geometrics, and abstracts — all elegant in their stark simplicity. 64pp. 8⅜ x 11. 0-486-44862-2

THE ARTHUR RACKHAM TREASURY: 86 Full-Color Illustrations, Arthur Rackham. Selected and Edited by Jeff A. Menges. A stunning treasury of 86 full-page plates span the famed English artist's career, from *Rip Van Winkle* (1905) to masterworks such as *Undine, A Midsummer Night's Dream,* and *Wind in the Willows* (1939). 96pp. 8⅜ x 11. 0-486-44685-9

THE AUTHENTIC GILBERT & SULLIVAN SONGBOOK, W. S. Gilbert and A. S. Sullivan. The most comprehensive collection available, this songbook includes selections from every one of Gilbert and Sullivan's light operas. Ninety-two numbers are presented uncut and unedited, and in their original keys. 410pp. 9 x 12. 0-486-23482-7

THE AWAKENING, Kate Chopin. First published in 1899, this controversial novel of a New Orleans wife's search for love outside a stifling marriage shocked readers. Today, it remains a first-rate narrative with superb characterization. New introductory Note. 128pp. 5³⁄₁₆ x 8¼. 0-486-27786-0

BASIC DRAWING, Louis Priscilla. Beginning with perspective, this commonsense manual progresses to the figure in movement, light and shade, anatomy, drapery, composition, trees and landscape, and outdoor sketching. Black-and-white illustrations throughout. 128pp. 8⅜ x 11. 0-486-45815-6

THE BATTLES THAT CHANGED HISTORY, Fletcher Pratt. Historian profiles 16 crucial conflicts, ancient to modern, that changed the course of Western civilization. Gripping accounts of battles led by Alexander the Great, Joan of Arc, Ulysses S. Grant, other commanders. 27 maps. 352pp. 5⅜ x 8½.   0-486-41129-X

BEETHOVEN'S LETTERS, Ludwig van Beethoven. Edited by Dr. A. C. Kalischer. Features 457 letters to fellow musicians, friends, greats, patrons, and literary men. Reveals musical thoughts, quirks of personality, insights, and daily events. Includes 15 plates. 410pp. 5⅜ x 8½.   0-486-22769-3

BERNICE BOBS HER HAIR AND OTHER STORIES, F. Scott Fitzgerald. This brilliant anthology includes 6 of Fitzgerald's most popular stories: "The Diamond as Big as the Ritz," the title tale, "The Offshore Pirate," "The Ice Palace," "The Jelly Bean," and "May Day." 176pp. 5⅜ x 8½.   0-486-47049-0

BESLER'S BOOK OF FLOWERS AND PLANTS: 73 Full-Color Plates from Hortus Eystettensis, 1613, Basilius Besler. Here is a selection of magnificent plates from the *Hortus Eystettensis,* which vividly illustrated and identified the plants, flowers, and trees that thrived in the legendary German garden at Eichstätt. 80pp. 8⅜ x 11.
                    0-486-46005-3

THE BOOK OF KELLS, Edited by Blanche Cirker. Painstakingly reproduced from a rare facsimile edition, this volume contains full-page decorations, portraits, illustrations, plus a sampling of textual leaves with exquisite calligraphy and ornamentation. 32 full-color illustrations. 32pp. 9⅜ x 12¼.   0-486-24345-1

THE BOOK OF THE CROSSBOW: With an Additional Section on Catapults and Other Siege Engines, Ralph Payne-Gallwey. Fascinating study traces history and use of crossbow as military and sporting weapon, from Middle Ages to modern times. Also covers related weapons: balistas, catapults, Turkish bows, more. Over 240 illustrations. 400pp. 7¼ x 10⅛.   0-486-28720-3

THE BUNGALOW BOOK: Floor Plans and Photos of 112 Houses, 1910, Henry L. Wilson. Here are 112 of the most popular and economic blueprints of the early 20th century — plus an illustration or photograph of each completed house. A wonderful time capsule that still offers a wealth of valuable insights. 160pp. 8⅜ x 11.
                    0-486-45104-6

THE CALL OF THE WILD, Jack London. A classic novel of adventure, drawn from London's own experiences as a Klondike adventurer, relating the story of a heroic dog caught in the brutal life of the Alaska Gold Rush. Note. 64pp. 5³⁄₁₆ x 8¼.
                    0-486-26472-6

CANDIDE, Voltaire. Edited by Francois-Marie Arouet. One of the world's great satires since its first publication in 1759. Witty, caustic skewering of romance, science, philosophy, religion, government — nearly all human ideals and institutions. 112pp. 5³⁄₁₆ x 8¼.   0-486-26689-3

CELEBRATED IN THEIR TIME: Photographic Portraits from the George Grantham Bain Collection, Edited by Amy Pastan. With an Introduction by Michael Carlebach. Remarkable portrait gallery features 112 rare images of Albert Einstein, Charlie Chaplin, the Wright Brothers, Henry Ford, and other luminaries from the worlds of politics, art, entertainment, and industry. 128pp. 8⅜ x 11.   0-486-46754-6

CHARIOTS FOR APOLLO: The NASA History of Manned Lunar Spacecraft to 1969, Courtney G. Brooks, James M. Grimwood, and Loyd S. Swenson, Jr. This illustrated history by a trio of experts is the definitive reference on the Apollo spacecraft and lunar modules. It traces the vehicles' design, development, and operation in space. More than 100 photographs and illustrations. 576pp. 6¾ x 9¼. 0-486-46756-2

A CHRISTMAS CAROL, Charles Dickens. This engrossing tale relates Ebenezer Scrooge's ghostly journeys through Christmases past, present, and future and his ultimate transformation from a harsh and grasping old miser to a charitable and compassionate human being. 80pp. 5³⁄₁₆ x 8¼. 0-486-26865-9

COMMON SENSE, Thomas Paine. First published in January of 1776, this highly influential landmark document clearly and persuasively argued for American separation from Great Britain and paved the way for the Declaration of Independence. 64pp. 5³⁄₁₆ x 8¼. 0-486-29602-4

THE COMPLETE SHORT STORIES OF OSCAR WILDE, Oscar Wilde. Complete texts of "The Happy Prince and Other Tales," "A House of Pomegranates," "Lord Arthur Savile's Crime and Other Stories," "Poems in Prose," and "The Portrait of Mr. W. H." 208pp. 5³⁄₁₆ x 8¼. 0-486-45216-6

COMPLETE SONNETS, William Shakespeare. Over 150 exquisite poems deal with love, friendship, the tyranny of time, beauty's evanescence, death, and other themes in language of remarkable power, precision, and beauty. Glossary of archaic terms. 80pp. 5³⁄₁₆ x 8¼. 0-486-26686-9

THE COUNT OF MONTE CRISTO: Abridged Edition, Alexandre Dumas. Falsely accused of treason, Edmond Dantès is imprisoned in the bleak Chateau d'If. After a hair-raising escape, he launches an elaborate plot to extract a bitter revenge against those who betrayed him. 448pp. 5³⁄₁₆ x 8¼. 0-486-45643-9

CRAFTSMAN BUNGALOWS: Designs from the Pacific Northwest, Yoho & Merritt. This reprint of a rare catalog, showcasing the charming simplicity and cozy style of Craftsman bungalows, is filled with photos of completed homes, plus floor plans and estimated costs. An indispensable resource for architects, historians, and illustrators. 112pp. 10 x 7. 0-486-46875-5

CRAFTSMAN BUNGALOWS: 59 Homes from "The Craftsman," Edited by Gustav Stickley. Best and most attractive designs from Arts and Crafts Movement publication — 1903–1916 — includes sketches, photographs of homes, floor plans, descriptive text. 128pp. 8¼ x 11. 0-486-25829-7

CRIME AND PUNISHMENT, Fyodor Dostoyevsky. Translated by Constance Garnett. Supreme masterpiece tells the story of Raskolnikov, a student tormented by his own thoughts after he murders an old woman. Overwhelmed by guilt and terror, he confesses and goes to prison. 480pp. 5³⁄₁₆ x 8¼. 0-486-41587-2

THE DECLARATION OF INDEPENDENCE AND OTHER GREAT DOCUMENTS OF AMERICAN HISTORY: 1775-1865, Edited by John Grafton. Thirteen compelling and influential documents: Henry's "Give Me Liberty or Give Me Death," Declaration of Independence, The Constitution, Washington's First Inaugural Address, The Monroe Doctrine, The Emancipation Proclamation, Gettysburg Address, more. 64pp. 5³⁄₁₆ x 8¼. 0-486-41124-9

THE DESERT AND THE SOWN: Travels in Palestine and Syria, Gertrude Bell. "The female Lawrence of Arabia," Gertrude Bell wrote captivating, perceptive accounts of her travels in the Middle East. This intriguing narrative, accompanied by 160 photos, traces her 1905 sojourn in Lebanon, Syria, and Palestine. 368pp. 5⅜ x 8½. 0-486-46876-3

A DOLL'S HOUSE, Henrik Ibsen. Ibsen's best-known play displays his genius for realistic prose drama. An expression of women's rights, the play climaxes when the central character, Nora, rejects a smothering marriage and life in "a doll's house." 80pp. 5³⁄₁₆ x 8¼. 0-486-27062-9

DOOMED SHIPS: Great Ocean Liner Disasters, William H. Miller, Jr. Nearly 200 photographs, many from private collections, highlight tales of some of the vessels whose pleasure cruises ended in catastrophe: the *Morro Castle, Normandie, Andrea Doria, Europa,* and many others. 128pp. 8⅞ x 11¾. 0-486-45366-9

THE DORÉ BIBLE ILLUSTRATIONS, Gustave Doré. Detailed plates from the Bible: the Creation scenes, Adam and Eve, horrifying visions of the Flood, the battle sequences with their monumental crowds, depictions of the life of Jesus, 241 plates in all. 241pp. 9 x 12. 0-486-23004-X

DRAWING DRAPERY FROM HEAD TO TOE, Cliff Young. Expert guidance on how to draw shirts, pants, skirts, gloves, hats, and coats on the human figure, including folds in relation to the body, pull and crush, action folds, creases, more. Over 200 drawings. 48pp. 8¼ x 11. 0-486-45591-2

DUBLINERS, James Joyce. A fine and accessible introduction to the work of one of the 20th century's most influential writers, this collection features 15 tales, including a masterpiece of the short-story genre, "The Dead." 160pp. 5³⁄₁₆ x 8¼.
0-486-26870-5

EASY-TO-MAKE POP-UPS, Joan Irvine. Illustrated by Barbara Reid. Dozens of wonderful ideas for three-dimensional paper fun — from holiday greeting cards with moving parts to a pop-up menagerie. Easy-to-follow, illustrated instructions for more than 30 projects. 299 black-and-white illustrations. 96pp. 8⅜ x 11.
0-486-44622-0

EASY-TO-MAKE STORYBOOK DOLLS: A "Novel" Approach to Cloth Dollmaking, Sherralyn St. Clair. Favorite fictional characters come alive in this unique beginner's dollmaking guide. Includes patterns for Pollyanna, Dorothy from *The Wonderful Wizard of Oz,* Mary of *The Secret Garden,* plus easy-to-follow instructions, 263 black-and-white illustrations, and an 8-page color insert. 112pp. 8¼ x 11. 0-486-47360-0

EINSTEIN'S ESSAYS IN SCIENCE, Albert Einstein. Speeches and essays in accessible, everyday language profile influential physicists such as Niels Bohr and Isaac Newton. They also explore areas of physics to which the author made major contributions. 128pp. 5 x 8. 0-486-47011-3

EL DORADO: Further Adventures of the Scarlet Pimpernel, Baroness Orczy. A popular sequel to *The Scarlet Pimpernel,* this suspenseful story recounts the Pimpernel's attempts to rescue the Dauphin from imprisonment during the French Revolution. An irresistible blend of intrigue, period detail, and vibrant characterizations. 352pp. 5³⁄₁₆ x 8¼. 0-486-44026-5

ELEGANT SMALL HOMES OF THE TWENTIES: 99 Designs from a Competition, Chicago Tribune. Nearly 100 designs for five- and six-room houses feature New England and Southern colonials, Normandy cottages, stately Italianate dwellings, and other fascinating snapshots of American domestic architecture of the 1920s. 112pp. 9 x 12. 0-486-46910-7

THE ELEMENTS OF STYLE: The Original Edition, William Strunk, Jr. This is the book that generations of writers have relied upon for timeless advice on grammar, diction, syntax, and other essentials. In concise terms, it identifies the principal requirements of proper style and common errors. 64pp. 5⅜ x 8½. 0-486-44798-7

THE ELUSIVE PIMPERNEL, Baroness Orczy. Robespierre's revolutionaries find their wicked schemes thwarted by the heroic Pimpernel — Sir Percival Blakeney. In this thrilling sequel, Chauvelin devises a plot to eliminate the Pimpernel and his wife. 272pp. 5³⁄₁₆ x 8¼. 0-486-45464-9

AN ENCYCLOPEDIA OF BATTLES: Accounts of Over 1,560 Battles from 1479 B.C. to the Present, David Eggenberger. Essential details of every major battle in recorded history from the first battle of Megiddo in 1479 B.C. to Grenada in 1984. List of battle maps. 99 illustrations. 544pp. 6½ x 9¼. 0-486-24913-1

ENCYCLOPEDIA OF EMBROIDERY STITCHES, INCLUDING CREWEL, Marion Nichols. Precise explanations and instructions, clearly illustrated, on how to work chain, back, cross, knotted, woven stitches, and many more — 178 in all, including Cable Outline, Whipped Satin, and Eyelet Buttonhole. Over 1400 illustrations. 219pp. 8⅜ x 11¼. 0-486-22929-7

ENTER JEEVES: 15 Early Stories, P. G. Wodehouse. Splendid collection contains first 8 stories featuring Bertie Wooster, the deliciously dim aristocrat and Jeeves, his brainy, imperturbable manservant. Also, the complete Reggie Pepper (Bertie's prototype) series. 288pp. 5⅜ x 8½. 0-486-29717-9

ERIC SLOANE'S AMERICA: Paintings in Oil, Michael Wigley. With a Foreword by Mimi Sloane. Eric Sloane's evocative oils of America's landscape and material culture shimmer with immense historical and nostalgic appeal. This original hardcover collection gathers nearly a hundred of his finest paintings, with subjects ranging from New England to the American Southwest. 128pp. 10⅝ x 9. 0-486-46525-X

ETHAN FROME, Edith Wharton. Classic story of wasted lives, set against a bleak New England background. Superbly delineated characters in a hauntingly grim tale of thwarted love. Considered by many to be Wharton's masterpiece. 96pp. 5$\frac{3}{16}$ x 8 ¼. 0-486-26690-7

THE EVERLASTING MAN, G. K. Chesterton. Chesterton's view of Christianity — as a blend of philosophy and mythology, satisfying intellect and spirit — applies to his brilliant book, which appeals to readers' heads as well as their hearts. 288pp. 5⅜ x 8½. 0-486-46036-3

THE FIELD AND FOREST HANDY BOOK, Daniel Beard. Written by a co-founder of the Boy Scouts, this appealing guide offers illustrated instructions for building kites, birdhouses, boats, igloos, and other fun projects, plus numerous helpful tips for campers. 448pp. 5$\frac{3}{16}$ x 8¼. 0-486-46191-2

FINDING YOUR WAY WITHOUT MAP OR COMPASS, Harold Gatty. Useful, instructive manual shows would-be explorers, hikers, bikers, scouts, sailors, and survivalists how to find their way outdoors by observing animals, weather patterns, shifting sands, and other elements of nature. 288pp. 5⅜ x 8½. 0-486-40613-X

FIRST FRENCH READER: A Beginner's Dual-Language Book, Edited and Translated by Stanley Appelbaum. This anthology introduces 50 legendary writers — Voltaire, Balzac, Baudelaire, Proust, more — through passages from *The Red and the Black, Les Misérables, Madame Bovary,* and other classics. Original French text plus English translation on facing pages. 240pp. 5⅜ x 8½. 0-486-46178-5

FIRST GERMAN READER: A Beginner's Dual-Language Book, Edited by Harry Steinhauer. Specially chosen for their power to evoke German life and culture, these short, simple readings include poems, stories, essays, and anecdotes by Goethe, Hesse, Heine, Schiller, and others. 224pp. 5⅜ x 8½. 0-486-46179-3

FIRST SPANISH READER: A Beginner's Dual-Language Book, Angel Flores. Delightful stories, other material based on works of Don Juan Manuel, Luis Taboada, Ricardo Palma, other noted writers. Complete faithful English translations on facing pages. Exercises. 176pp. 5⅜ x 8½. 0-486-25810-6

FIVE ACRES AND INDEPENDENCE, Maurice G. Kains. Great back-to-the-land classic explains basics of self-sufficient farming. The one book to get. 95 illustrations. 397pp. 5⅜ x 8½. 0-486-20974-1

FLAGG'S SMALL HOUSES: Their Economic Design and Construction, 1922, Ernest Flagg. Although most famous for his skyscrapers, Flagg was also a proponent of the well-designed single-family dwelling. His classic treatise features innovations that save space, materials, and cost. 526 illustrations. 160pp. 9⅜ x 12¼. 0-486-45197-6

FLATLAND: A Romance of Many Dimensions, Edwin A. Abbott. Classic of science (and mathematical) fiction — charmingly illustrated by the author — describes the adventures of A. Square, a resident of Flatland, in Spaceland (three dimensions), Lineland (one dimension), and Pointland (no dimensions). 96pp. 5³⁄₁₆ x 8¼. 0-486-27263-X

FRANKENSTEIN, Mary Shelley. The story of Victor Frankenstein's monstrous creation and the havoc it caused has enthralled generations of readers and inspired countless writers of horror and suspense. With the author's own 1831 introduction. 176pp. 5³⁄₁₆ x 8¼. 0-486-28211-2

THE GARGOYLE BOOK: 572 Examples from Gothic Architecture, Lester Burbank Bridaham. Dispelling the conventional wisdom that French Gothic architectural flourishes were born of despair or gloom, Bridaham reveals the whimsical nature of these creations and the ingenious artisans who made them. 572 illustrations. 224pp. 8⅜ x 11. 0-486-44754-5

THE GIFT OF THE MAGI AND OTHER SHORT STORIES, O. Henry. Sixteen captivating stories by one of America's most popular storytellers. Included are such classics as "The Gift of the Magi," "The Last Leaf," and "The Ransom of Red Chief." Publisher's Note. 96pp. 5³⁄₁₆ x 8¼. 0-486-27061-0

THE GOETHE TREASURY: Selected Prose and Poetry, Johann Wolfgang von Goethe. Edited, Selected, and with an Introduction by Thomas Mann. In addition to his lyric poetry, Goethe wrote travel sketches, autobiographical studies, essays, letters, and proverbs in rhyme and prose. This collection presents outstanding examples from each genre. 368pp. 5⅜ x 8½. 0-486-44780-4

GREAT EXPECTATIONS, Charles Dickens. Orphaned Pip is apprenticed to the dirty work of the forge but dreams of becoming a gentleman — and one day finds himself in possession of "great expectations." Dickens' finest novel. 400pp. 5³⁄₁₆ x 8¼. 0-486-41586-4

GREAT WRITERS ON THE ART OF FICTION: From Mark Twain to Joyce Carol Oates, Edited by James Daley. An indispensable source of advice and inspiration, this anthology features essays by Henry James, Kate Chopin, Willa Cather, Sinclair Lewis, Jack London, Raymond Chandler, Raymond Carver, Eudora Welty, and Kurt Vonnegut, Jr. 192pp. 5⅜ x 8½. 0-486-45128-3

HAMLET, William Shakespeare. The quintessential Shakespearean tragedy, whose highly charged confrontations and anguished soliloquies probe depths of human feeling rarely sounded in any art. Reprinted from an authoritative British edition complete with illuminating footnotes. 128pp. 5³⁄₁₆ x 8¼. 0-486-27278-8

THE HAUNTED HOUSE, Charles Dickens. A Yuletide gathering in an eerie country retreat provides the backdrop for Dickens and his friends — including Elizabeth Gaskell and Wilkie Collins — who take turns spinning supernatural yarns. 144pp. 5⅜ x 8½. 0-486-46309-5

Browse over 9,000 books at www.doverpublications.com